THE
HAND

THE
HAND

EMEKA OKEANI

1603 Capitol Ave., Suite 310 Cheyenne, Wyoming USA 82001
1-888-980-6523 | admin@urlinkpublishing.com

URLink Print and Media is committed to excellence in the publishing industry.

Published in the United States of America

Library of Congress Control Number: 2021920017
ISBN 978-1-64753-972-6 (Paperback)
ISBN 978-1-64753-973-3 (Digital)

13.09.21

Surviving and thriving in your new land
and
Understanding and harnessing
the new entrant potential

TABLE OF CONTENTS

DEDICATION

This book is dedicated to my late father, Pa Johnson Okeani, who remains an eternal inspiration and support in my endeavors. He always encouraged us, his children, to shoot for the moon.

Thank you, Father, for your words of wisdom and the parental nudge to pursue our dreams.

May you continue to rest in peace.

"Context is to data and information what water is to a dolphin."
-Dan Simmons

PREFACE

As you read this book, please bear in mind that the author maintains certain premises. These stances provide the background for this book. Besides, they will assist the reader in perhaps understanding the contextual framework of the book.

The first is that humans are naturally nomadic. Therefore, migration remains a part of the human condition and history. Immigration has been beneficial to the United States of America in general. Today, 10 percent of all residents in the United States are foreign-born. The Pew Research Center reported 44 million people living in the United States in 2017 were born in another country. Come to think of it, the original pilgrims were immigrants who left Great Britain searching for a more conducive environment, community, and society. They set out to the new land and eventually cobbled what is now the United States of America. Since then, herds of immigrants have arrived at America's shores. Here is the crux; I refer to legal immigration, whereby potential immigrants engage in a formalized and government-approved migration. Throughout this book, whenever reference is made to immigration, it is that described above.

Studies and data have mostly pointed to the advantages that continue to accrue to the nation from legal migration. Please, observe the following names from various walks of life; the sciences, arts, business, journalism, sports, music, politics, literature, etc., who have contributed towards our great society. These names represent only a tiny fraction of immigrants to the United States but ones that you may be familiar with. Albert Einstein, Arnold Schwarzenegger, Sofia Vergara, Bob Hope, Alex Trebek, Bruce Willis, Gloria Estefan, Jackie Chan, Sammy Sosa, Madeleine Albright, Elon Musk, Audrey Hepburn, Levi Strauss, Henry Kissinger, Bob Marley, Rupert Murdoch, Andrew Carnegie, Michael J. Fox, Chinua Achebe, John Muir, Melania Trump, Jim Carrey, Joseph Pulitzer, Patrick Ewing, Eddie Van Halen, Mario Andretti, Yo-Yo Ma, Oscar de la Renta, Saint Patrick, Steve Job's father who emigrated from Syria, and many others. These names have something in common; they emigrated from another land to the United States. From Ronald Reagan to Donald Trump's grandparents or great-grandparents, or great-great-grandparents, all of the recent past American Presidents were immigrants to America. Presidents Ronald Reagan, Bill Clinton, and Obama share Irish ancestry from at least one of their forebearers. The Bushes possess English and German descent, Obama's paternal ancestry traces back to Kenya, Africa, and Donald Trump's heritage was of German and Scottish immigrants to America. All with immigrants' ancestry. From a global perspective, consider that Prince Phillip of England (the Duke of Edinburgh) was born in Greece, emigrated to the United Kingdom, and became a naturalized British citizen in 1947. It does not get any higher than that in the citizen hierarchy ladder. How about these migrations: famous actress Audrey Hepburn from Belgium to the United Kingdom; world-renowned soccer player Leo Messi from Argentina to Spain; actor Jonny Depp from Kentucky to France; iconic singer Tina Turner from

Tennessee to Switzerland, and music performer Nicki Minaj from Trinidad and Tobago to the United States?

In the past two decades, about 25 percent of all venture capital start-ups were launched by immigrants, including eBay, Google, and Yahoo. Forty percent of recent Silicon Valley innovations originated from immigrants. According to a 2011 report by Forbes, 40 percent of all fortune 500 companies were started by immigrants or their children. It is noteworthy that those who contribute towards our country's advancement are not always necessarily born within its borders.

The second premise is that, generally speaking, most immigrants' odyssey tends to be similar, from conception, arrival, culture shock, adjustment, and eventual assimilation. There is usually a psychological and mental gyration that occurs. These upheavals can be minor or pronounced depending on the individual and the specific new environment. Sometimes, this shake-up can be immediate or delayed, but it happens as a rite of passage to the new arrivals' journey.

In the new arrival's experience, an oft-misunderstood notion centers on the speed of acculturation. Some citizens might be frustrated at a lower-paced integration of the new entrant. An individual is an embodiment of their background, life experiences, and culture. Such grounding, lens, and perspective are ingrained and a part of the being. Shedding this deeply entrenched modus operandi can be challenging. Imagine Bob born and raised in the United States where there is a lesser intrusion in its citizenry's private lives, comparatively speaking. Bob migrates to an authoritarian society with unchecked government mandates and control of its citizen's lives or a nation with stringent firearms restrictions for its populace. We remind ourselves what social scientists have been drumming for centuries, the relativism of culture. In other words, culture tends to be relative to its surrounding. Therefore, one cannot judge another culture based on one's cultural underpinnings.

So, for the new entrant, an expectation of a quick cultural flip is unrealistic and a heavy burden on the new arrival. Interestingly, the bicultural tug on the immigrant can be nerve-racking and arduous. The author discusses this phenomenon and its consequences throughout this book. Sometimes, this metamorphosis can be misread as not being sufficiently loyal to the immigrant's newly adopted country. This process can last for a few years, decades, and even a generation. The transformation depends on the individual, the level of immersion in one's new land's culture, and the degree of abandonment or modification of the old country's culture.

Globalization and technology have become the game changers on a host of things, including human migration. Through technology, digital devices, and social media, the world has become smaller and more intimate. Platforms such as Facebook, Instagram, YouTube, Twitter, Snapchat, WhatsApp, Tik Tok, etc., enable communication and information sharing in more widespread ways and in an instant. By so doing, cultures are exposed, learned, and emulated. Gone are the days when the primitive tribes at the end of the earth are not known or studied. Today, for example, grandmothers and grandfathers can and do communicate with their grandchildren they have never physically met. They interact with them via video or audio devices. Undoubtedly, such ease of communication and electronic bonding have lessened the immigrant's physical connectivity burden with those they left behind.

Invariably, the acceptance and condoning of children and relatives' migration to another land are more permissible and tolerable. Nowadays, a great-grandmother seated at her porch in Africa can facetime with her great-grandchild in Chicago, Atlanta, or Los Angeles. Entertainment icon Kim Kardashian's Instagram post is watched and enjoyed instantly by the teenagers in Warsaw or Rio de Janeiro. Fashion trendsetter Givenchy's latest style in France is immediately consumed on Fifth Avenue New York, or London's New Bond Street. A sixteen-year-old in Belarus lip syncs the latest Beyonce or Billie Eilish song on Tik Tok. An aspiring viral comedian meshes funny memes on YouTube from his enclave in Mexico City and instantly played worldwide. One can argue that through technology and globalization, cultures are diluting and morphing, at least, at the fringes. Within some world cultures, some chasm is emerging between the generational cohorts. While physically residing in a particular country, the younger cohorts embrace some cultures of different societies transmitted and internalized through technological devices and platforms. This inclination towards external cultures is usually at the displeasure of the older generational cohorts who lament the waning of traditional and embedded values, culture, and identity.

Whereas the newcomers are initially accomodated and tolerated at the arrival and early acculturation stage, they must quickly learn and shed some outlier norms, beliefs, and practices that are glaringly inconsistent with their new land's culture. For example, migrating from a society that is unaccepting and disrespectful of female bosses by male workers is untenable in American society. Clinging to such anachronistic and un-American perspective runs afoul of your new community. The onus equally lies on the newcomer to step up their assimilation antennae and process. There are no free rides.

As you read this book, allow these premises to play in the background for you. They will help enrich your understanding of the author's assertions. Also, some sections might not directly relate to your experience. For example, native-born readers should exercise patience as they read some aspects of the newcomer's prior life and experiences before emigrating. Those previous stories weave a wholesome tapestry for the reader. They provide the backdrop for the newcomer's present situation and circumstance. Because this book might be read by native-born and newcomers, a glossary is provided at the end to assist the reader in fully understanding the words and terms. Finally, all names used in this book do not represent the individuals' actual names cited and the locations were intentionally scrambled. However, the events, occurrences, and activities mentioned are real.

"Adventure is the child of courage."
- Jonathan Huie

CHAPTER 1

GOING WEST! (PREPARING FOR THE JOURNEY)

No one who immigrates to another country wakes up one day and suddenly decides to emigrate. It is a process. It is a path that starts mentally, fueled by three or four main propellers: economics, adventure, escape from political or religious persecution, or love and romance. The economic thrust stems from wanting to better one's life and calculus that seeking that golden fleece is better elsewhere than one's current place of abode. This search for a greener pasture is pursued through education or job opportunities. Besides, the knack for adventure can lure one into emigrating to another land. Usually, this aspiration is materialized by possessing sufficient income or funds to fulfill such a dream. Also, political or religious persecution compels migration to another land. It, therefore, becomes migration compelled by circumstances and desire for survival.

Another trigger is love and romance, which often entail following a loved one or being dragged by a loved one. I use pulled because some budding romance requires some cajoling to be uprooted from one's original home country. Such a waltz requires the inviter to promise a better environment and society. A new combined life that the male or female seducer paints as elevated, jazzy, and freer. They send a few glamorous pictures of beautiful and exquisite locations to whet the enticed appetite further. Only selected solo pictures or ones with only a couple of friends or relations are sent to magnify the seeming desperation and loneliness. It would be detrimental and counterproductive to send party scene pictures and celebrative gatherings. The batting rate for such coaxing's success is usually high because the ambers of the romance that existed before departure still smolder. Over time, the smoldering wanes and vanishes.

The initial thoughts start to crystallize into a call for personal action. Actions are truncated into three main categories; sourcing the avenues to pursue one's goal, identifying the country or location of interest, and activating the required network towards making the dream a reality. Regarding education, schools and universities are researched for financial admission requirements. Evaluations of the surfaced options are made regarding affordability and methods of payments. Any known personal contacts and networks are revived to obtain practical feedback on life in the proposed country. Where applicable, the region, city, and community are examined for a better fit with one's background. Funding sources are evaluated and determined. Is one responsible for the financial expectations, or are other family members to play some role in bearing the burden? It switches into a communal undertaking in societies that practice and embrace an extended family-type kinship. The intended traveler starts to craft their story and pitch to the rest of the family network delivered individually or in a group setting. Under this communal system, the traveler promises to assist other extended family members upon attaining financial stability. Committed obligations could range from sending specific amounts yearly to the kinship purse or paying the younger member's secondary and post-secondary tuition. Often, these understandings and obligations are never written in a contract. Instead, it is a civic and moral commitment that is not explicitly named. Some societal practice limits the traveler's funding source to the nuclear family. Others involve sponsorship by the community. Nonetheless, the point here is the communal or nuclear family rallying support for the proposed emigration.

A self-funding traveler would personally organize their funds accordingly in pursuing the goal at hand. For a working or job seeking traveler, efforts are geared towards winding down from current job and aligning activities to migration. Overall, the traveler starts preparing those to be left behind. More on this later. Should the worker possess some network of individuals, usually forerunners who are now settled in the new country, communication with them intensifies. This connection is crucial because of the role this network plays in the welcoming, onboarding, and eventual resettlement in the new country. Most successfully integrated immigrants credit the on-ground support system. This informal infrastructure enables the latest entrant to step into the new environment surrounded by individuals whose culture and behavior, though somewhat modified, are closer to what they are accustomed to. This phenomenon is universal, whether one is migrating from Europe, Asia, Africa, South America, or anywhere in the world. This clustering behavior is why ethnic enclaves dot our cities, towns, and communities. This clamor to congregate around familiar people and culture remains an inert human survival instinct. We tend to gravitate towards similar people, especially from a cultural and eco-social standpoint. Whereas many advantages accrue from this practice, it sabotages interethnic interactions and learnings. One could argue that a part of the racism experienced in societies can be attributed to ethnic isolations that deprive communities of adequate and robust exchanges and interactions. Just look around the city, town, or the metropolis where you live, and you will find very entrenched ethnic neighborhoods with their unique subcultures. Some of these communities can look like garrisons where an "outsider" is viewed with suspicion and caution. Have you ever had an experience where you were looked at as someone who probably lost their way when approaching a strictly ethnic location or marketplace? If you have been spared of such experience, you are one of the rare lucky folks in America. But for the fun of it, go try it out. I bet you know exactly where to go to observe this learning encounter. Better yet, waltz into a different ethnic neighborhood church during Sunday worship, and you will experience this educational blessing. While some have downplayed the importance of neighborhoods in urban settings, there is general agreement that ethnic neighborhoods remain a crucial settlement and adaptation mainstay for new arrivals. These enclaves often meet newcomers' needs and expectations in family ties, familiar culture, affordable housing, and help in seeking and securing work. At the very least, these enclaves represent a vital bridge for the newcomer in transitioning from the cultures of the old country to the dominant American culture. It has been argued that while some newcomers might and do move to more affluent neighborhoods once their economic lot improves, others remain to continue perpetuating their ethnic identity. The latter group resists complete assimilation and acculturation into the mainstream culture. Consider the following neighborhoods, and one may appreciate their importance: Little Havana section of Miami,

Chinatown in Flushing New York, Greektown in Baltimore Maryland, Swedish Andersonville in Chicago, German Fredericksburg enclave in Texas, Danish Solvang area in California, Little Bombay section in Northern New Jersey, Little Persia neighborhood in Los Angeles, Little Ethiopia in Washington D.C., Little Brazil in South Framingham Massachusetts, Little Saigon in Philadelphia, Jewish Borough Park in Brooklyn New York, and so on. We will reserve a robust debate about the benefits and disadvantages of ethnic neighborhoods for social scientists, who have continued to research and write on the subject.

Back to the traveler. This time, the job seeker. Expertise and skills are revisited, dusted off, and ready to sell to whoever is prepared to buy. Opportunities are sought through the internet, print media, and word of mouth. Applications are completed, follow-up interviews (nowadays via an electronic medium) are conducted, and decisions are made. Then comes the process of filing the proper papers with the Immigration and Naturalization Service, Homeland Security, etc., to seek the needed traveling papers to emigrate. The traveler's anxiety level is now heightened to almost a fever-pitch level. The primary reason for such worry is that now everyone around you knows you are working towards relocating to another country. You have now placed all your eggs in one proverbial basket. You cannot fail now. If you are a student, all your friends know that you will be leaving soon; they are gradually tuning you out of their existence. Understandably, they must do this to mitigate the blow your absence may cause them. It is a survival move. If you are employed, your job, as a matter of fact, no longer includes you in the internal memorandums and emails. Your church, mosque, or synagogue family remember you in prayer every week for the intended journey.

Meanwhile, you have not yet obtained your traveling visa and papers, so you are on a pressure cooker. A refusal to issue traveling papers to you or an outright rejection of your visa application spells shame and possible ridicule. To successfully plan for emigration, you must focus tenaciously on the effort and invariably detach from friends and community as you travel domestically within your country of origin, gathering the required documents. Family members who have contributed funds towards your endeavor are now expecting you to travel soon. You asked and yearned for it. Now, grab it; they would nudge.

I have explained this stressful process to illustrate the initial anxiety potential immigrants experience even before setting foot in their soon-to-be adopted country. Ask any immigrant, and they will share the agonizing process, either inflicted by the bureaucracy of their native country or the scrutiny and vetting of the United States government agencies, or both. These are necessary steps and procedures, but they remain vexing and arduous, nonetheless. Sleepless nights and endless worrying become the daily routine of the intended traveler. You are now consumed by the objective of obtaining the required papers for travel.

You are then scheduled for your interview for the visa and travel documents. At this juncture, if your blood pressure is suspect before now, it kicks into high levels, your ulcer starts to act up, and you begin to have bad dreams. Dreams of rejection at your interview, or being sent back to gather more papers, etc. You become very religious quickly (if you were marginally religious), attending church service every Sunday and asking for prayers. You become inwardly focused, shunning your friends and sometimes your relations. The degrees of intensity for the activities mentioned above may vary among individuals. Still, there is no arguing these phenomena occur regardless of which corner of the earth one emigrates to. These events demonstrate the trials and trepidations of the aspiring immigrant. The ups and downs of emotions and the upheaval inherent in the migration process and journey start to wear one down. It is not for the faint-hearted. Some aspirants abandon their plans and goal after a first significant setback or obstacle. Tenacity and focus are required to navigate the process and eventually achieve success. You ultimately obtain the necessary papers and visas to travel. You become relieved that this portion of the ordeal is over. You commence your travel preparations by purchasing your travel tickets and items you consider essential for your journey.

Your emotions start to play tricks on you by planting some doubt in your mind if you are making the right decision, as this move tremendously alters your life's trajectory forever. Your current abode and circumstance are known entities. Wherever you are going is unknown and untested by you. You are gripped with fear and uncertainty. You shake off the jitters and attempt to focus on the positive. You justify your decision by recounting that you have been trying to make some headway with your life unsuccessfully for several years. You cement your decision by recounting stories of some friends or relatives who ventured out and succeeded. By golly, you will succeed as well, you reassure yourself. You have now cleared all the cobwebs and mentally ready to plunge. The time had finally arrived. Some relations and friends see you off and wave safe journeys and goodbyes. And off you go!

While en route, you contemplate different scenarios that could unfold. What if this entire endeavor becomes a bust? What if you cannot find your footing in the new land? What if things go well? You engage in normal human suppositions and calculus. Your meandering mind is on autopilot. You settle your nerves by falling asleep. You wake up, and your mind and emotions are still actively at work. Now, it is unto all your friends, some lifelong, and relations that you would no longer physically see, at least for many years. Your eyes well up with tears. You think about the aged relatives who, through a natural cycle, may die before you could see them again. You wonder if you had adequately fulfilled your obligations to them while in your old country. You take some cursory inventory of your relationship with all you interacted with until now. You catch yourself from drifting into a mushy mind territory. You fling yourself back to reality and the here and now. Next thing you know, you are in your new land.

While each immigrant's circumstances might vary from another, certain commonalities exist. The rallying by the family, both morally and financially, in supporting the adventure, the anxieties associated with the journey, the emotional tug of leaving loved ones behind, the consternations of such life-altering endeavors usually manifest in the process. The levels or the degree of each element, while different, are ever-present.

*"Culture is the widening of
the mind and spirit."*
- Jawaharlal Nehru

CHAPTER 2

TOUCH DOWN!

(Arrival and Initial Awakenings)

Arriving by plane, the wheels touch down on the tarmac, and the crew announces the arrival to the specific city in your new land – America. If by sea, the boat or ship docks at an American shore. If by automobile, you cross the border into your new land. All of this carries a mixed emotion of joy and fear. Gladness, because you have accomplished the first significant step in your ultimate life's journey. Scared because the future is uncertain. Very uncertain. You have made a gamble, and pray to be correct. You have cast the dice of your life. The frightfulness of this endeavor suddenly hits you. The enormity of your chance play dawns on you and, for a moment, freezes you.

You start to take in your new environment but not entirely because you are temporarily numbed by the stark reality of life here and now. Though you had planned things in your mind and maybe even on paper, the rubber is currently meeting the road, and you do not have a return ticket. This is it! You asked for it, and now it is on your plate, all garnished. Ask any immigrant of their first moment of arrival, and most will recall the specific mixed emotion of joy and fear. The apprehension is because of the uncertainties that lie ahead. Yes, some may have relatives and friends meet them upon arrival, but their minds were racing with other emotions other than seeing relatives' or friends' faces. You are anxiety-laden, you are nervous, you are happy, you are overwhelmed, you are perhaps exhausted, and you are partially paralyzed by fear – the fear of the unknown and the future.

You immediately observe and notice things, items, and activities that make America unique, even at the airport, seaport, or highway. These things are usually not apparent to the citizens. Some are subtle, while others are pronounced. A coworker named Victor, who emigrated from Eastern Europe, relayed that his first experience with America occurred at the airport when a relative, who had come to pick him up, treated him to French fries. "French fries!" I sarcastically exclaimed. He quickly retorted, "You don't understand, ever since I was a growing boy in my native country, I had dreamt of eating the real American French fries, and I was partaking of one of my life's dream …sticking authentic American fries in my mouth, unbelievable!" His emotional outburst and expression drove the point home for me. It was more than French fries. It was a culture; it was a new life; it was everything progressive and futuristic. Victor still enjoys his French fries but has curtailed his portions and their consumption frequency. He is now into fitness, weight loss, dieting, and health consciousness culture.

How about the immigrant Cecelia from Africa who marveled and was enthralled by the motorists' orderliness obeying unattended traffic lights? She explained that on her first day of arrival, she kept watching the traffic lights and the automobile drivers' lawful behaviors that her driver asked her why she kept starring at the traffic lights and road. She unhesitatingly exclaimed, "Oh my God, I cannot believe how diligently drivers are obeying these signs and lights without a police officer standing at every corner; this would never occur orderly back home. Motorists, motorcyclists, and pedestrians are so unruly back home that they would never obey these lights," she concluded. Further down the drive, she continued by advancing that one of the root causes for their behaviors was extreme poverty. She ranted how only a few citizens were corruptly diverting the nation's wealth for their families' enjoyment. Therefore, she suggested that the rest of the citizenry register their disgust and frustration through mass disobedience and flouting established laws. She jokingly asked the driver how a hungry and angry person could honestly obey an artificial contraption such as a light signal? Of course, her question was tongue in cheek. However, her fulmination was anchored in the populace's mass deprivation in her old country. People could never reconcile the abundant natural resources with the abject squalor that many citizens endure. Upon arrival, Bolade, also from Africa, was utterly amazed that one could purchase an item, return it another day and receive a full refund without a fuss from the store. More on this later.

Think of Juan from South America, who could not understand the iris identification system at his entry airport. When asked to place his forehead on the gadget for his iris reading, he became so shaken because he feared something puncturing his eyes while opening them in the apparatus. He would attempt to place his forehead against the instrument bar but could not keep steady. The U.S. Immigration and Customs officer eventually assisted Juan in properly setting his head and opening his eyes for a good iris read. This experience amazed Juan that it left an indelible mark on him. Juan, today, is a very successful digital devices developer in California. He credited the described airport arrival episode with influencing his education and career choice.

I have to tell the story of Tabita, who had emigrated from a rural South American community. Even in her old country, Tabita's exposure to the big city was limited. When she arrived in the United States and was taken to a Super Walmart store, Tabita was fascinated by the orderliness of the products-display aisles and the sheer number of available items. She unconsciously found herself in the middle of the store gazing around in amazement and awe. Her relative, who had taken her to the store, purposely did not interfere with this unique experience for Tabita. The relative just stood on the side while Tabita was taking it all in. Other shoppers would glance at Tabita and kept walking. Some just shrugged their shoulders in wonder. What they did not realize was that the experience was one-of-a-kind for Tabita. The citizens are used to this perceived orderliness, neatness, and abundance. To Tabita, it was an out-of-body experience. She later confessed to her relative that she had never physically seen so much food and varied consumer items in one location. Never! For many months while she lived with her relative, she pleaded to tag along for any trip to any store, grocery, mall, or corner gas station. One could easily be perplexed how such trivial and mundane circumstances could stir such a huge emotional impression on an individual. It is all about our backgrounds, exposures, surroundings, and perspectives. What one takes for granted could be viewed and weighed heavily by another. As the old saying goes, "one man's trash is another man's treasure." To this day, Tabita has not erased this vivid and lifelong mark of her first few days in her new land. She often tells the story with a reflective undertone.

First impressions on arrival in a new land matter. They stay and stick with you. One hopes they are good impressions. Sometimes, they are not all positive. Unfortunately, you cannot quickly and easily shed them either. Diego, who emigrated from Chile, recounted that he was rudely ridiculed for not fully speaking or understanding English upon his arrival. He lamented that though he could manage some minor conversational English, he could not speak fluent English to hold a good conversation. Diego recounted his experience: As an Immigration and Customs agent was processing him, he was asked to present some documents. He had managed to shake his head and pointed to his ear that he could not understand what the agent said. He could read the agent's body language as he had started to be irritated while calling out for a Spanish interpreter. Unfortunately, none was readily available since their flight emanated from Chile with a majority of Spanish-speaking passengers. This situation tied up all of the Spanish interpreters at that moment. While awaiting any available interpreter, the agent again attempted to communicate with Diego to no avail. It was then that the agent, assuming that Diego was not capable of understanding any English at all, blurted "Stupid and dumb" to Diego's hearing.

You know what is often said …. People tend to learn derogatory words in a language first. Diego knew what those two uncomplimentary words meant. Indeed, they were hurtful. Diego could not do much other than taking the ridicule and verbal abuse stoically. After all, he was on the verge of fulfilling his life's dream, coming to America. He would not risk it due to what he had considered a minor insult. However, the experience left a mental scare on him. Though Diego speaks very fluent English today, he initially suffered a harsh reality in his new and adopted country. I will discuss this further in chapter 5.

The new arrival starts to observe and venture into society, most likely starting with members of their ethnic community where available. Ethnic communities pulsate with more life and vibrancy than the mainstream community, at least in the newcomer's eyes and psyche -- the activities representing an accurate and exacting replica of the old country. For example, the genre and style of music played by area stores are familiar tunes. The food and cuisine odors filtering into the olfactory organs come with nostalgic smells. The language spoken made the environment seem like the newcomer never left the old country. Even the attires worn by many reflected the assemblages donned at home. It all brought memories to the newcomer without physically going back to the old country.

A visiting relative from an Eastern European old country observed to his host after spending a few hours at their ethnic neighborhood that "It felt exactly like home." He further commented, "That's why you guys don't visit often. You have every bit of home here." According to his host, he was right. He explained that whenever newcomers felt homesick, they would intentionally spend some hours at their bustling ethnic enclave and receive their "home inoculation" that gave them cover for a while. Another immigrant once proclaimed that it was a ritual for his entire family that included children and grandparents to visit and indulge in ethnic interactions (shopping and eateries) every fortnight at the ethnic neighborhood central point near downtown. He submitted that such immersions outweighed oral history or book knowledge. Their children experienced their heritage culture firsthand and in real time.

Within this enclave, something often subtle but sometimes conspicuous happens. There usually are two worlds in these ethnic communities. Some choose to remain firmly rooted in their ethnicity. Others shed their ethnic community connections and quickly immerse themselves in the mainstream culture. Since their arrival, these individuals may have attained higher financial and educational levels and have detached and self-isolated from the greater community. They may still maintain very few friends in the community, but they are "gone" and "AWOL" from mingling and associating with their erstwhile friends and relations. I resist passing judgment on this behavior and stance. Allow me, though, to amplify a comment by a member of an ethnic community. She stated that those who distance themselves from the community return when they desire ethnic participation in weddings, graduation, child christening, or even funerals.

Interestingly, some ethnic community avoidance stems from unhealthy behaviors and attitudes such as competition, jealousy, and hatred meted on each other. These negative biases result from each other's intimate knowledge, which perpetrates this attitude at higher levels than in a typical societal setting. Immigrants who hail from a particular country always remember the commonality of heritage. This common thread among them exposes the human instinct of comparison and judgment. They mindlessly compare their children, jobs, affluence, and marriages. As they engage in these diminutive and relationship-busting behaviors, they become crass and vindictive. You may think that I am exaggerating this phenomenon and dynamic. I challenge you to inquire about this destructive disposition within the various ethnic enclaves. In the eighties, a college mate, who later joined the Immigration and Naturalization Service, told me that their most significant source of personal immigration violations came from ethnic communities. Initially, they would jump to pursue and investigate provided information, which sometimes was anonymous, until they discovered that most of the time, such materials were nothing but vengeful counterpunches from within their communities. They often correlated violation reports with individuals experiencing upward mobility in the community. For example, when one purchased a house, a new car, attained a higher educational degree, landed a good job, or gained a career promotion. The envious, spiteful, and highly competitive ethnic associates tended to attempt to trip their seemingly more successful members up by presenting them as targets for the Immigration and Naturalization Service. They had been privy to some original inconsistencies in the newcomer's earlier story to the government agencies during their legalization process. They, therefore, fling those oddities to derail their rivals. Perhaps, a thorough scientific study is warranted to determine and decipher the human emotions and attributes that drive this universally practiced behavior among ethnic communities.

As the new entrant starts to step into the new environment and community, certain aspects of the society and culture shock the newcomer. When I first arrived in the United States more than four decades ago, I was disturbed by the concept of relegating seniors to assisted living and nursing homes. To me then, it represented the abandonment of parents and abdication of care by selfish children, who could not bear the inconveniences of their old and perhaps frail parents. To understand my mindset, the culture that I was raised in, and whence I came from rewarded parents for their efforts and sacrifice in raising their children by openly bringing them back into the kids' abode when they got older. People took pride in caring for their parents by assimilating the older parents into their often busy lives. This practice was an unwritten contract and obligation that passes from one generation to the next. So, I immediately surmised the self-centeredness of children towards their aged parents. I was one-dimensional and shallow in my view because I did not consider that my old country's culture did not possess or provide assisted living or nursing home facilities. Therefore, when parents became old and could not adequately cater for themselves, there is no option but to absorb them into the children's living arrangements.

It took a sociology class during my undergraduate education, deeper thinking, and the understanding of cultural relativism to shed the negative notion I had about parents and nursing homes. Unfortunately, I was viewing my new cultural setting from the lens of my cultural background. This ethnocentrism is common with immigrants and only gets adjusted and refined after educational exposure, learning experiences, and sheer hard knocks. I would submit that this nagging frame of mind and subsequent behavior impairs and delays the newcomer's more rapid acculturation into his or her new land. As simple and easy as it may seem, it isn't easy to quickly jettison the cultures and socialization ingrained in the recent entrant. Remember, our identities and self-esteem center around our cultural underpinnings. To instantly let go of them without first building on one's adopted culture leaves one rudderless without any cultural anchor. So, the newcomers cling tenaciously to what they knew and what makes them who they are. Do you blame anyone?

Another example of cultural relativism and enduring values is Indian arranged marriages. Up to a whopping ninety percent of Indian marriages are arranged. No, this is not a typographical error. Even westernized Indians living in the

U.S. and European countries choose this over a free-choice one. A 2013 IPSOS survey found seventy-four percent of young Indians (18-35 years old) prefer arranged marriages over free-choice marriage. These arranged marriages' divorce rate is not runaway, as they hover around one to one and a half percent. This rate compares to a U.S. divorce rate of approximately forty-five percent. The point here is not to engage in scientific analysis and comparison. Instead, to illustrate, while some cultures may view arranged marriages as weird or unbecoming, it seems to work in the Indian cultural context. For example, under the Indian setting, processes exist to express expectations and desired attributes of a future life partner, usually to older relatives. A vetting process, which includes examining and evaluating the families, is performed. An introductory procedure, a meeting of the would-be partners, and follow-up initiated. The entire process could last from a few days to one year or beyond, depending on the potential partners' sense of urgency. Without any formal and written template or rule book, this process is usually carried out without a hitch. It is an ingrained cultural pillar in the lives of Indian citizenry. Often, when one migrates to the United States from India, vestiges of this practice and value persist. An Indian American recently equated the Indian arranged marriage system to the online dating platforms, with familiar humans replacing artificial electronic devices as intermediaries. Interesting! Chapter 10 holds a more robust discussion on cultural relativism.

Along the line of cultural relativism, consider Andrew, who emigrated from a society where the economic system was not credit-dependent. In Andrews' society and culture, if you crave something, you must endure until you save enough money to acquire the goods or services. Upon emigrating to the United States, Andrew could not fathom himself buying things on credit. He often was irritated at correspondence and solicitations for acceptance of credit offers from banks or other lending institutions. Every car that he purchased was paid for in cash, though he drove a car until it was ready for the junkyard. Andrew today is well assimilated into the American credit economy. The purchase of a house did it for him. By not being able to save for a house, he succumbed to his adopted country's economic model. Ironically, he sometimes laments that he could have leveraged the financial credit system long ago when he realized the benefits of the model. He had latched on to his original background much too long on this matter, he confessed. Again, it would not have been easy for Andrew to immediately embrace and practice the credit and consumerism system without losing who he was. This conflict illustrates the typical conundrum that confronts and challenges the newcomer -- the ability or inability to access the American credit system properly and proportionately without excesses and overindulgence. Easing into the consumerism model and orientation must be carefully orchestrated and mindfully initiated not to be carried away by its simplicity and access. Whether one likes it or not, your new land's credit system is inescapable, though. Be familiar with this significant element of this financial system, the "credit score." A credit score represents a consumer's creditworthiness. It is based on credit history, which encompasses debt, repayment history, and financial obligations fulfillment behaviors. Financial institutions use the credit score to evaluate and determine a potential borrower's probability to repay loans on time.

A credit score significantly impacts one's financial life. It represents a key factor in a lender's decision to offer credit. It also permeates other aspects of one's universe. For example, insurers use credit scores to set premiums for automobile and homeowner's coverage. Landlords use them to evaluate potential renters, financial houses, and other organizations to determine the required initial deposit to obtain smartphones, utilities, or cable services. Simply put, it can be a lever or a hammer. Good scores elevate and can be leveraged, while bad scores pound you into missing out or paying more. I hate to put it this bluntly; The credit score is our "branding" societally speaking.

Somewhat parallel to the credit economic model is the concept and practice of returning purchased goods and obtaining a full refund (most of the time) should you not be delighted with the items. To some of the newcomers, it is surprisingly an alien concept. In some societies, once a purchase is concluded, the transaction is complete, and no returns are allowed or permissible. Bolade was mouth-agape when he changed his mind about the color and style and returned the pair with a full refund after purchasing a pair of shoes. Before returning the pair of shoes, Bolade had rehearsed his reason to provide to the store clerk. Much to his surprise, the clerk did not even query him on the return. She examined Bolade's tendered receipt, pulled open her cash drawer, and counted the exact original payment amount unto Bolade's outstretched hand. As he walked out of the store, Bolade joyfully and gleefully mumbled to himself, "This is truly God's own country."

In Bolade's old country, once you paid for a good or service and walked away even for a minute, you could not return the item or service. Never, no matter how cogent your reason was, such as size or color. Should the initially purchased size not fit, you are expected to buy a different size that fits. The onus wholly and squarely rested on the buyer to make the correct choice once. Bolade relayed how he had bought a pair of pants for his high school graduation. He had purchased this at a market store where there were no facilities to try them before payment. You were supposed to know your measurements without considering the different sizing and style of cuts by various brands. He paid for his normal-sized pair. Bolade got home and tried the pair and immediately observed that he could not fit into them comfortably. He rushed back to the market store to obtain a larger-sized pair, only to be callously told by the same store owner who had sold him the item just forty-five minutes earlier that he could not exchange the pair for a different size. The store owner's skewed argument and rationalization were that they had already consummated the earlier transaction and, therefore, was a closed matter, period. If he needed another size, he had to pay for a new pair. Dejected and disappointed, Bolade went home. Unfortunately, this was typical in transactions in Bolade's old country, he explained. So, one could see why Bolade was jubilant when offered a full refund for a pair of shoes that he had bought, taken home for some days, and returned. That represented nirvana to Bolade.

In 1977, when a new arrival from Belize accompanied a relative to a drive-in movie theater, he could not believe his eyes. The newcomer had never seen nor experienced anything like this. Drive-in theaters are now relics of the past, but they represented popular pastime spots in their hay days. For the benefit of the newer generation cohorts, these theaters were locations where goers viewed movies from their vehicle's privacy and comfort. Vehicles are parked to permit a view of a large screen. Soundtracks are broadcast to car radio stereo systems. This recreational feature in America was popular, especially in rural and suburban areas. Parents and adults could watch movies with their infants and younger children in tow. The air of informality surrounding this setting was appealing to families. Besides, it was a hit for dating. Technological advancements in home TV and media outlets led to the demise of the drive-in theater niche. Nonetheless, this entertainment offering represented the American persona of freedom, privacy, and ruggedness. The new arrival was awestruck by this unique relaxation feature in his new land. He remembers and recounts this experience as if it were yesterday.

Food is not only for nourishment but also represents our identity. We embody the food we eat, familiar with, and are exposed to. Though globalization has mitigated the availability of any cuisine anywhere globally, it can sometimes present some challenges. Yes, if you look hard, you can find any ethnic food in most cities. But we are not talking about eating out daily or the like. We refer to the average Joe or Jane finding the right grocery stores to purchase hitherto familiar foodstuffs. One can easily ask, why not go to the nearby fastfood joints and eat the good old burgers or fries, or meatloaf, or potato and gravy dishes. Not that simple. The newcomer's taste buds and palate must adjust to these new foods. An immigrant from Micronesia once told me that it took him almost eight months to begin to enjoy a good old hamburger. It was not for lack of trying to eat them. He claimed that his taste buds could not quickly adjust to these new foods. Whenever he consumed these burgers, they were tasteless to him. Not being a scientist, I could not respond to his story's veracity.

Nonetheless, the point here is that even food that one eats in their new land undergoes adjustment. To some, this food adaptation could be shorter than others. A few European immigrants have told me that though the essential ingredients may be similar to American cuisine, the preparations and eventual taste and palatability are different.

Take Jack's case, who emigrated from England roughly twenty-eight years ago. He claimed that he could not taste and enjoy the good old American potato for about two years after arriving in the United States. The potato was one of Jack's favorite food, even in his old country. So, upon arrival, he picked up where he left off with potato, attempting to consume it as he regularly did back home. Much to his chagrin and disappointment, the American potato did not taste the same as his old country's potato, he claimed. First, he thought it was the brand and type of potato he was buying or ordering at restaurants. He tried the whole gambit of potatoes he could find at grocery stores or served at eating places. No luck. They were all almost tasteless, regardless of what condiments garnished them. Then after about nine months, he began to enjoy the taste somewhat, and another six months saw an enhanced taste that equaled the taste of potatoes he was used to in his old country. He had wondered about the experienced phenomenon and was told that his buds were adjusting to his new food offerings in his new land. He claimed that most food items, chicken, meat, fish, and even other vegetables, were subjected to the same palatal adjustments and modifications.

One of the thriving businesses in ethnic communities is the grocery and supermarket stores. These stores act as a bridge for the newcomer in adapting to their new environment, including food consumption. Take a look around, and you will find these stores dotting various neighborhoods. Interestingly, they are immune to big grocery chain takeovers. They serve a particular niche and cannot be adequately replicated by big grocery chains. Sociologically, these stores represent gathering holes for members of the communities. They also serve as advertising avenues for the communities. Real estate posters, hair salon cards, tax preparation, insurance service advertisements, etc., are usually displayed in and around the stores. Furthermore, these stores represent an effective word-of-mouth platform for information dissemination. Some budding relationships and romances have even been traced to chance encounters at these ethnic stores.

"*A house builder does not stop building because a nail broke. They change the nail.*"
-African proverb

CHAPTER 3

HOUSTON, WE HAVE
A PROBLEM!

The daily dilemma of new versus old cultures does not abate. It accelerates as one engages in one's new land. The newcomer faces a plethora of challenges. Confronting and sorting these tensions remains the only practical path to integrating into one's new environment. Those that retreat from dealing and resolving these conflicts prolong their assimilation into their new society. The initial feeling of befuddlement and even disorientation, often termed culture shock, must be tamed for survival in the new land. The stark reality of this path that the newcomer must navigate causes a cry out, "Houston, we have a problem." The individual's "Houston" most often is one's inner tussle involving heart, head, strength, weakness, willpower, and fortitude.

If you are a Christian, you hearken to your inner bible verses of Mathew 9:14-17, Mark 2:21-22, and Luke 5:33-39 that admonishes about not putting new wine into old bottles; else the new wine will burst the bottles and be spilled. Though you remember in your bible lessons that Jesus was using these items and images to address the Pharisees' inquiries on seemingly conflicting practices, you fleetingly ponder if such teachings apply to your circumstance.

Let us explore and examine what some of these problems might be.

If a student, it suddenly dawns on the newcomer that sustaining one's education requires self-efforts by obtaining a job to augment whatever is sent from the family. This work must be performed while enrolled in studies. How can I juggle these and continue to garner good grades at school? One wonders. In the meantime, in the old country, America's notion as a progressive nation where hard work and personal effort are rewarded dominate the mindset. So much so that any funding from the old country is begrudgingly provided. After a few years at best, such financial support withers and forever withdrawn. You are now purely on your own and must make it happen. Houston, we have a problem!

As a student, you start your classes and notice your nice-looking classmates. You are lumped into groups for your class assignments and some friendships develop. These innocent friendships begin to morph into deeper relationships that you are not ready to entertain. You are here to study, period! You reiterate to yourself. Sorry, the web is reeling you in without your even realizing it. You freeze because of this new tension of bouncing hormones and your studies' focus. What is going on here? You continually ask yourself. You vow not to succumb to this rushing train of romantic feelings. You become stoic and reticent to the other party hoping that feigning disinterest would shoo away the feeling. You are not successful. You and your potential lover are being drawn together for the wrong reasons; you articulate. But the mindless passion train has picked up speed instead. Oh no, Houston, we have a problem!

The compounding conflicts and activities cause you to minimize and sometimes cut off communication with your family that supported your emigration. They start to worry about you and ask anyone they know in your new land if they have seen you lately. These fellow residents convey these inquiries to you, and you get irritated by their nosing around and withdraw even further. In the meantime, that "friend" you worked so hard to avoid and run from has magically gotten closer to you. You wonder how that happened, but it is too late now. Suddenly, the looming financial hardship makes you jump at the teasing of a proposal to move in together with your friend that you are now inseparable from. Within weeks, both of you are signing rental documents to live together. You wonder if it was true love or fear of the impending personal economic difficulties that drove your recent move-in decision. You will never determine the answer, but it nags at you constantly. The answer is that various circumstances squeeze you, a newcomer to make unanticipated decisions. You are being dealt a hand. A hand of unavoidable situations that confluence at the same time. Your brain tells you that you are engaging in this hurried "shacking up" stupidly and recklessly. Still, your amygdala (emotion-processing limbic system) has now hijacked your mental faculties, hence the rash decisions.

Your newcomer's mind starts to get warped when your live-in friend hints at the relationship's seriousness. How do you spell "possible marriage"? You freak out because that is the last thing on your mind, if ever. You are not ready to even entertain such a thought. But it is here before you could think about it. The path for such a life journey is usually not neat and methodical. It springs on you, especially at an ill-prepared moment. If you are a student, you reason that you are yet to achieve your reason for migrating to a new land and if you are a worker, you rationalize that you needed to establish a firmer foundation before this romantic bliss. Fear, concern, and anxiety overwhelm you, and it starts to show with your irrational decisions, heightened irritability, and extreme withdrawal. In the meantime, your "friend" has been observing and absorbing these behavior and attitudinal wheels coming off. Houston, we have a problem!

The dynamics of conflict and confusion prevail upon you, the individual, whether a student or a worker. The reality of falling in love versus the goal for one's emigration tugs relentlessly at the new entrant. After all, you are human, and loving someone is an inert human condition. You cannot successfully suppress it even if you tried. You are torn because this was not even in your playbook when you conceived your emigration plan to a new land. How could you? However, did you think that you will remain perpetually your age when you migrated or stay single the rest of your life? The point here is that while planning your migration, you never think that far ahead of your life. Your primary focus is getting there and achieving your first big-ticket item, be it education or a job. However, life is not that neatly lived. Things, events, activities, etc., come at their own pace and time. You must grapple with them and hope and pray for successful outcomes. The new entrants' experiences may not be drastically different from their native counterparts, except that events, activities, and circumstances that might plague a native for an extended period get compressed for the immigrant upon arrival and entry into their new society. Ask any immigrant, and they will tell you of periods when they thought that they might have offended God because of the aggregation of challenges almost simultaneously. The truth is that it is necessary for the filtering that must occur to transition to the new culture. The newcomers are put under the proverbial fire to make them stronger and more adaptive to their new land. It is a rite of passage. Undergoing this forging is usually difficult but in the end, it adequately prepares you to not only engage in your new land but adjusts your paradigm to what you must possess to live and thrive.

Problems emerge from different angles and avenues. You find yourself isolated and even depressed. You are very new in the new land and have not yet established and nurtured a supportive network of friends and acquaintances. You perhaps do not even know where to start in cultivating friendships. To you, everyone seems to be mired in their individual lives and matters. Due to your perception, which was mostly flawed, you retreat to your shell, thereby causing more isolation from the community. Houston, we have a problem!

Finances and making ends meet present their challenges. While you are now enjoying all the trappings of an organized and advanced society, it is not without its price. The monthly bills for your enhanced lifestyle start to mount. Nothing is free in your new land; you murmur to yourself. And this takes me to the bewilderment expressed by a newcomer in the early eighties. He had arrived at the Seattle/Tacoma airport, popularly known as the Sea-Tac airport. He became biologically pressed and therefore needed to ease himself. Upon discovering the toilets through the airport signages, he was astonished that he had to deposit some token amount to access the stalls. "Unbelievable!" He exclaimed. "Paying to ease me?" Eventually, he found some coins in his wallet to drop in the designated slots for access. He quickly surmised that money drives everything in his new land. Houston, we have a problem!

A friend told a story of the early years of emigrating to the United States. He was attending college in Arizona along with a childhood friend. During spring break, they decided to visit another childhood friend in Iowa. My friend and his schoolmate had not embraced and fully integrated into the American economic system. They had not established enough credentials to obtain credit cards. They conducted all of their transactions in cash as in their old country. They had planned to make the drive leisurely and stress-free. So, they planned on staying at a hotel about halfway along the way, going and returning. Then the shocker came when they pulled into the parking lot of a hotel. They went into the lobby and to the counter with their luggage in tow. They inquired if rooms were available. Upon an affirmative answer, the clerk requested their credit cards. They had none. The clerk advised that a credit card was essential for renting rooms. They proudly boasted that they had the raw cash to pay for their stay. The clerk reiterated that their policy required a credit card in one's name. "This hotel chain is just too picky and uppity," they reasoned. So, they stormed out and drove down the road to another lesser brand hotel. No luck. Stack reality hit when they begrudgingly went to a low-end hotel and were equally denied renting rooms due to an absence of a credit card. Houston, we have a problem! Upon realizing their inability to secure rooms without this vital economic and security tool: credit card, they continued their journey without any hotel stay.

An experience almost akin to the above had happened to Albert. As a newcomer, Albert had sought to rent an apartment in Dallas, Texas, shortly after arriving in the United States. Upon inquiry, he was directed to an apartment agency who showed him several apartments, and he chose one. The apartment manager requested some documents and information from Albert, including prior residence, past utility bills, and references. Albert was stunned. He did not possess this information and documents. Albert had not lived in the country long enough to have established these vital profile data. Remember, he just emigrated to the country. Even after Albert explained his newness to the country, he was rejected by the apartment entity and could not rent. Houston, we have a problem!

The following story occurred in the Northeastern Region of the United States. The wife of a Fortune 500 company executive recounted this experience. She worked at an organization that interacts with clients daily. A client had picked up a casual conversation with her. Their discussion led to mentioning that her husband was usually out of town for work on weekdays. Without hesitation, the client asserted, "Oh, your husband must be a truck driver." She was temporarily immobilized by shock and disgust simultaneously for this client's supposition and conclusion. You see, the wife in this story and husband had emigrated from Africa. So, to the client, who was Caucasian, a traveling newcomer from Africa must be a truck driver. Little did the client know that the supposed truck driver was a bona fide national executive of a major American corporation. Of course, the executive's wife would not let the conversation end on that note. After she composed herself, she quietly but proudly explained her husband's position and title, the "Vice President" in this household name company. The client mustered a subdued, "Wow, that is great!" and scrammed. Houston, we have a problem!

The above described encounters and experiences are not confined to workplaces, college campuses, and by chance alone. They sometimes occur in elementary school settings, as naïve and innocent such an environment might deceitfully pose. How about this story for an illustration? A pupil of an assimilated immigrant came home to tell what took place in her classroom that day. She wore a shirt with an often sold brand at a particular boutique store in their neighborhood. This burly boy could not contain his curiosity. So, he approached her within the teacher's earshot and inquired if she was wearing the brand name or a fake? The girl quickly shot back by stating it was 100% authentic. That was a little too much for the boy because he probed further, "Which store did you purchase it at?" She would not bow, and she pridefully named the store. He could not take it, so he dug deeper by stating, "It could not be because you folks are poor and cannot afford to shop there." Yes, in the middle of an elementary school of six and seven-year-old children. At that point, the teacher who heard all the exchanges intervened and disciplined the boy accordingly. Houston, we have a problem!

One can easily dismiss this childlike encounter as naivete and simple innocence of kids. Except that on a closer examination, it was more significant than just children interacting in the classroom. After all, how could a six or seven-year-old have formed a perspective and worldview that people different from him, especially dark-colored, were usually poor and could not enjoy the largesse of society as he and his family and the likes could? Where did he learn that it was acceptable to belittle a classmate of a different race and ethnicity? What gave him the audacity to challenge a classmate for indulging in what is available to everyone? Where did the intolerance emanate from? Let us leave those ponderings to social scientists. The point here is simply that as a newcomer, your children are never immune to the knocks and digs that accompany your newness and acculturation in your new land. Recognize the bumps and traps littered everywhere, and choose to navigate them cleverly, mindfully, and purposefully.

Did you think that perhaps grandparents are spared? Think again! They are not the least exempted from matters of intolerance and mistreatment. Stephan's entire family-children, parents, and grandparents- emigrated to the United States under a program of resettling former allies whose territory had been overtaken by hostile forces and regime. Stephan's grandfather was a highly ranked army officer in their old country. He had adamantly refused to emigrate, vowing to die in his homeland. After intense emotional persuasions from family members, including his grandchildren, he reluctantly agreed to join the rest of the extended family in emigrating to the United States. As expected, he presented the most challenge in adapting to their new land. Remember, in their old country, he was a supremely privileged man that enjoyed all the trappings of a highly ranked military officer. All of those privileges vanished overnight, as he became just an ordinary person in his new land. He struggled tremendously with this new reality.

One afternoon, as all the grandchildren had gone to school, and his son and his wife were both at work, he decided to take a leisurely walk around the neighborhood. He walked and felt it was a good exercise for him, so he continued to walk back and forth along the neighborhood's long and winding trail. Then, a police car with some blue turning lights and no siren pulled up beside him. The officer alighted from the vehicle and respectfully asked for his information, name, address, etc. He froze with fear and could not fully articulate his thoughts and answers. The officer sensing his disorientation, started to calm him down, explaining that someone in the neighborhood had called the police that some strange man seemed to be casing the neighborhood. Inferring that he may have been termed a likely thief or loiterer, at best, he broke into tears. He nervously narrated to the officer that he was a high-ranking army officer in his old country and had risked his family member's lives assisting the United States in executing the proxy war in his home country. Being assumed or termed a thief was the farthest expectation he had hoped for emigrating to the United States. He was inconsolable for weeks, even by his son and the entire family. Houston, we have a problem!

Recently, it was reported in the news that a real estate agent in Brooklyn, New York, was caught on video berating an Asian gym manager to "go back to China." The agent allegedly refused to wear a mask inside the gym as required because of the pandemic. As the manager reminded him to wear his mask, the agent threw his mask on the floor and screamed at him with the statement. Another gym user captured the incident with his mobile phone and posted it on Tik Tok. The insulted manager is a bona fide American citizen. Assuming that the agent is an American citizen, his rights and privileges as a citizen are the same as the gym manager. No more, no less. Though the manager might have emigrated from another land, he is now a full-fledged American. Period. His status allows him to enjoy all the entitlements associated with citizenship. Houston, we have a problem!

A wealthy middle eastern family had sent their daughter to attend a Boston, Massachusetts University in the early '90s. Like the rest of the entire family, their daughter was surrounded by maids, housekeepers, chauffeurs, and other house help in their native country. She had never lived alone before emigrating to the United States. Of course, in Boston, there were no housekeepers and helpers to cater to her needs. She was lost. Attending to basic needs such as grocery shopping became very tedious for her. Houston, we have a problem! In later years, she confessed that she was depressed, lonely, and helpless when she first arrived in her new land. Ultimately, you want your initial cry out of "Houston, we have a problem" to become "Houston, problem resolved."

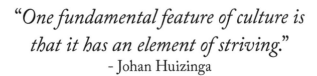

"One fundamental feature of culture is that it has an element of striving."
- Johan Huizinga

CHAPTER 4

READY, SET, GO (THE REAL JOURNEY BEGINS)

The minor cobwebs of the initial challenges and reality check are cleared. The real journey now commences. Yes, the voyage begins. The proverbial rubber is now meeting the road. It is noteworthy that the separation of the new entrants that will survive and thrive and those that will succumb to the lure of life temptations in the new land start to occur at this juncture. The latter group begins to fall prey to illegal drug use, fraudulent activities, and criminal activities. They are sluggards that aspire to ascend to middle class and affluent ranks at an accelerated rate. They have concluded that slogging or "paying their dues" through hard work is unnecessary. They can equally attain the American dream through their crooked paths, they reason. Little do they know that their new society's law enforcement dragnets and legal systems are far more robust, complex, and expansive than what they left behind in their old countries. They believe they are clever and hardened enough by hardship to outwit the American system. They have abandoned all the bulwark against mischief and criminality and instead chosen the lassitude of deviance and illegality. They often start small with petty crimes and fraudulent activities and graduate to master fraudsters and scammers with international connections and reach. They believe they can drive the fancy cars without sacrificing a lot or purchasing that envied mansion effortlessly or date the prettiest female at the nightclubs with their ostentatious spending and exaggerated display of wealth.

Interestingly, visit with them after five years, and they would most likely be begging for loose change to tide them over. I venture to submit that I have not yet met one who chose the path of criminality or con artistry to more quickly accumulate wealth that had been sustainable over a long period. Nothing is sweeter than a well-earned good reputation, riches, excellent health, and supportive family anchored in whatever religion or faith you practice. Anything else is fleeting and superficial. This lesson must be ingrained in each newcomer's psyche and thinking. Moreover, the law enforcement web in your new land is much more complex, sophisticated, and innovative than you can imagine. Therefore, the only proper avenue to success and thriving in America is the straight and clean way. Most immigrants who have achieved the so-called American dream always kept their noses clean and to the grindstone in their endeavors.

After being jolted by the initial brushes of reality, those who have chosen the law-abiding, honest, and steadfast route are now determined to make hay while the sun is shining. For students, they have ultimately settled on their degree or vocational path and eventual career goal. They become laser focused on accomplishing their academic pursuits. They find and keep a part-time job either at their institution or nearby organizations. The job engagement would augment the dwindling funds from their families. They quickly learn to spend conservatively, especially in items such as apartment room sharing, carpooling where necessary, and public transportation utilization. Those that hitherto could not cook become amateur cooks overnight and gradually perfect that competency. They must summon all these survival skills and tactics and deploy them masterfully.

The individuals who emigrated for jobs apply similar approaches to fundamental survival in the areas described. However, a jobholder's avenue towards economic emancipation takes a speedier route. For example, they start to step into the credit economy and dip their toes into building credit through their employment. Establishing creditworthiness within the American economic system, after all, is no longer alien to the newcomer. Indeed, it is a necessity for solid financial grounding. Observe that creditworthiness does not automatically translate to credit spending like a drunken sailor. Instead, it represents the capacity to flex it when needed to advance one's economic lot. The American financial worthiness and prestige are viewed from the lens of funds at hand or in the bank, hard assets, and credit level one can attract. This pathway is the holy grail that the economic system is built on. There is no escaping this cornerstone in survival and thriving in this society. That is the hand that you are dealt with as a new member of the community. You quickly become familiar with the question, "cash or credit" as you shop.

Let us discuss another commonality as you fully engage in your new society. The degree of adjustment for all newcomers to this experience varies. It largely depends on the practice in the society you migrated from. Some cultures' traditions with this are closer to the American attitude while other cultures are divergent from the American experience. What is this, you ask? The individualism penchant. This phenomenon is starker when you arrive from communities where the silos between individuals are not ironclad. Allow me to explain what I mean here. First, I am in no way referring to the economic system where the destiny of the community is hinged together. Capitalism is the norm and unquestionably the American system. No argument there. I am referring to the practice where individuals generally tend to live their lives in compartmental ways.

Imagine living in an Apartment complex or subdivision or condominiums, and you hardly know or seem to know your immediate neighbors. Yes, you may shake your head in disagreement that you know all individuals and families in your complex or neighborhood. After all, you play cards with Mary or Joe that live around the corner. You go to sports events and games with Stan and Jessica, who are just next door. I doff my hat to you. However, you are in the minority. Have you ever visited a friend, family, coworker, or acquaintance, and you casually ask, "Who is your neighbor?". First, you get a befuddled look as if you just crossed the line of decency and veering into the nosy territory. The answer comes, "I don't know, we rarely see each other, and when we do, we are both in a hurry coming or going." As if you surely want to commit a friendship suicide, you probe further, "How long have you lived here?". Now you get a get-out-of-my-business look with an answer "Oh, about five years." Think about it, for about five good years, you have not taken the time to know your neighbor? This indifference is not an indicting statement because I practiced the same behavior many years ago. I am illustrating how society's pressures compress our ability and, some would argue, the capacity to interact and mingle as we are caught up in our individualisms.

It is advanced that part of the reasons for individual isolation lies in the safety concern. I once asked a law enforcement officer acquaintance why he thinks people are not as socially mingling to the extent possible. He looked at me as if to say, are you seriously asking this question. He then replied, "Safety, man!" People do not want to be hurt, and there are many crazy folks around, so people stay to themselves or only stick with the people they know well. The newcomer must understand and appreciate this dynamic of individual isolation. Self-preservation is paramount to citizens and takes precedence over socialization and the building of friendships.

The other individualism that I reserved until now is the rugged individualism that is a major pillar of Americanism. One could safely argue that America was conceived, built, and nurtured on individual responsibility and rewards. This underpinning is not at the exclusion of other relevant virtues. Most, if not all voluntary immigrants, starting from those on the Mayflower to those that landed on Ellis Island, to those that came after them are driven by sheer personal responsibility and doggedness. They realize and understand that America provides the best environment and opportunity for individual success with personal efforts and hard work. Yes, even with all its societal flaws, America represents the topmost nation on earth where personal endeavor, sacrifice, and drive rewards abundantly. Of course, the playing field is usually not all level for all people all of the time. Ask any immigrant, though, starting from those that emigrated since the seventeenth century; you are likely to hear that America captured their hopes, aspirations, and desires compared to whence they migrated from. This hopefulness is why newcomers would endure or tolerate discriminatory behaviors meted on them. This endurance is because it pales with the poverty, lack of opportunity, hopelessness, and despair prevalent in their old country.

Sophia and Thomas, a wife and husband duo, emigrated to Virginia in the nineties. They searched for a place to live and were referred to an apartment and home rental referral agency. An agent took them to a neighborhood home for their viewing and evaluation. While inside the then-vacant house, the agent felt obligated to admonish the couple that the area of town was known for the racial and ethnic intolerance of the inhabitants. The agent had expected the couple to be taken aback by such revelation. Instead, they asked, first by Sophia, "Would they come into our house to express their distaste for our living here?" The agent caught herself stumbling for an answer due to the couple's lackadaisical attitude. She summoned, "No, of course." The couple then commented, this time, by Thomas that "As long as they do not come into our house to register their ignorance, they could care less of what their minds harbor." The agent immediately thought that this couple would survive and thrive in their new land with such an attitude.

You see, Sophia and Thomas had just emigrated from a very oppressive country with an ironclad regime, where personal freedoms are a rarity. Even consumer items were rationed, and choices were limited. So, to them, the occasional encounter of racism or the like were just minor annoyances when compared to their erstwhile experiences. As long as they did not get knocks at their door from the State Security Service, life was cozy. Those dreaded 2 am surprise visits from the State "law enforcement" officers remained nightmarish. With that gone, anything is bearable and tolerable, they reasoned.

Sophia and Thomas's story had a happy ending twist to it. When they eventually moved into their home, neighbors surprisingly met them with warmth and friendliness. In fact, on the third day of their move-in, their neighbor to the right knocked on their door. Their initial reaction was, "Oh no, here they come, the intolerant and ignorant folks." As Thomas slowly opened the door, a couple handed them a plate of freshly baked cookies with a handwritten note that read "Welcome to the neighborhood." With mouth agape, Thomas received the gift and offered a subdued thank-you. The next day, a family, husband, wife, and two children came to their door and extended a heartfelt welcome; at least that was Sophia and Thomas's impression. For the next several weeks, neighbors showered them with a seemingly joyous welcome. So, they became reflective and somewhat perplexed. This perceptual conflict was because the housing agent's picture and perspective were incongruent with the unfolding events. Was the agent the real problem, they wondered. One thing was certain, they lived in this neighborhood for four and a half years and never felt discriminated against, at least openly, and never experienced any hostile or intolerant behavior towards them. They could never decipher the initial attitude and feedback provided by the agent. Once the agent completed her task of assisting the couple in securing a home, interactions with the couple ceased. The lesson they took away from that experience was that they must form their impressions and opinions. Allowing others' views to infiltrate their space was bound to skew or muddy it up. Sophia and Thomas became lifelong friends with some of their neighbors at the subdivision.

Unfortunately, this next story had a sad ending, but it illustrates the treacherous and sometimes dangerous terrain the newcomer negotiates in the quest to assimilate in a new land. At times, the path is littered with the booby traps of life in a new world that could turn tragic. Adolphus was big and burly at six feet three inches and weighed about two hundred and fifty pounds. He liberally utilized the gym lifting weights and therefore was healthy and fit. Adolphus emigrated to the United States in nineteen eighty-five and chose to settle where his childhood friend Patrick lived in Memphis, Tennessee. After about a week of rest from his long trek to America, he started to take in the environment, sights, and scenes.

One fateful day, he had accompanied his childhood friend, who was temporarily harboring him to a neighborhood store. A 7-Eleven store, to be exact. Trouble began when upon alighting from their car and approaching the store, about four young adults were loitering near the entrance. Adolphus observed them and inquired of Patrick what these "small chaps" were doing when they should be at their homes. You see, Adolphus was already passing judgment on these young adults based on his background. In his old country, young folks this age ought not to be wasting their precious time hanging out in the vicinity of a store. They could be more useful to themselves and their parents at home. Patrick, who had lived in his new land for four years, hushed Adolphus to mind his own business, explaining that in this country, the young adults could partake of whatever pleases them as long as they were not infringing on anyone else's right or space. Adolphus heard Patrick's admonition, but it never sunk in.

As they approached the door, Adolphus's long arm accidentally brushed one of the young adult's sides. The young adult protested angrily. As Adolphus was paralyzed for a second on the young adult's vehement disapproval and was tongue-tied, Patrick immediately apologized on his behalf, and they went into the store. While inside the store, Patrick was pressed and went to use the restroom. Almost immediately, the aggrieved young adult stepped into the store to buy some item. According to the store clerk, the young adult continued to berate Adolphus for harshly brushing against him while entering the store. At that point, Adolphus felt that this tiny boy must not insult him unnecessarily. He had now become irritated that this young chap, as he called them, was becoming a nuisance to him. Adolphus grabbed the young man by the collar of his shirt as if to knot a tie and shook him by the head and neck to desist from his harassment. He did not end there. He proceeded to utter what ultimately became his "death sentence" by saying to the young man, "If you continue, I will handle and harm you mercilessly." He then let go of the shirt-collar grab. Towards the end of his utterance to the young man, Patrick came out of the restroom and heard Adolphus's words. As Patrick continued to pick up items from the shelves and Adolphus was tagging beside him, Patrick commenced his piece of advice to Adolphus. He explained that what could be termed a simple threat in their old country with no real repercussion could carry a more profound connotation in their new land. He cautioned Adolphus to refrain from using words such as "I will harm or hurt you," etc. Little did Patrick know that his statements and admonitions were an ominous prediction of what was to unfold.

The young man had felt severely threatened by Adolphus and had expressed fear to his colleagues outside. With herd-mentality, they decided to harm Adolphus before he got any of them. Remember that with Adolphus's background and his relative newness to America, he was completely done with that episode and moved on. Not so with the young man and now with the whole gang. The aggrieved young man was determined to strike back at Adolphus, whom he felt had intimidated him due to his size. Lo and behold, one of the young adults exclaimed that he had his gun with him and handed it to the aggrieved young man. It did not help matters that they were smoking marijuana outside the store. Adolphus and Patrick had no earthly idea what awaited them outside. While waiting, the aggrieved young man begged for some more weed, which his friend readily provided. After a few hard puffs and inhales, he signaled he was ready. Ready to do battle with the bully, they reasoned.

Unsuspectingly, Adolphus and Patrick completed their shopping and headed for the door. It was late evening, and darkness had started to descend. Adolphus was carrying a bag on his left arm as they walked out and towards their car. The aggrieved young man suddenly appeared and fired two shots towards Adolphus at a close range. He did not miss. Adolphus fell straight to the ground of the parking lot. The young man immediately ran towards a waiting car full of his friends. Shaking and confused, Patrick was able to run back to the store to request they call 911. Remember, those were the times when cell phones were not commonplace.

Thus, ended the life of Adolphus. Two cultures collided head-on. The misunderstanding and misreading of cultures, to a large extent, led to the tragic outcome of their encounter. The newcomer's journey includes rapidly learning and appreciating your new land's basic and rudimentary practices and ways of life. Adolphus paid dearly for the lack of a thorough understanding of certain behaviors, words, and moves in his new land.

"Ignorance is darker than the night."
-African Proverb

CHAPTER 5

OH NO, FLAT TIRE! THE STARK REALITY HITS

(Paradox of America)

You are now on slow-speed cruise control in your new land, coasting along on your American journey. Things and activities hitherto oblivious to you start to beckon in more pronounced fashions. You choose to ignore them, but they scream louder, almost hurting your eardrums. You must now listen, you must pay attention, you must now see if only for self-preservation. Your new land rife with abundant opportunities also bears some ugly sides. Racism hits the newcomer. Sometimes, it is subtle. Other times, it is blatant. In case you are wondering if I am sinking into the abyss of delusion and unfounded imaginations, allow me to share some stories and experiences recounted to me over the years.

Katarina, who emigrated from Eastern Europe and trained as a health worker was employed by a home healthcare agency. Like other employees, Katarina was sent to area homes to attend to mostly elderly clients in service of minor healthcare needs. She arrived at the intended house and was let in. Once the client saw and heard Katarina's Eastern European accent, the client vehemently protested that he did not want to be touched or attended by a non-American worker. He immediately called the service agency and demanded that Katarina leave his premises and a genuine and red-blooded American sent in her stead. No need to bore you with the eventual resolution, if you call it that. Katarina was made to leave and not attend to the elderly and ailing client. So, for a moment, consider yourself in Katarina's shoes. Here is a well-trained healthcare worker ready to deploy her knowledge and skills on a client who abjectly rejected her services because of her nation of origin and ethnicity.

One more story about home healthcare challenges. An owner of a thriving home healthcare agency told how a Caucasian friend quietly begged her to send an American-born Caucasian worker to nurse her ailing father. According to the proprietor, the requester stated, *"You know how old people are, and my father fits that mold. Therefore, please note that if you send a non-American-born worker, my father's therapeutic exercise will be in vain. He will not focus on getting well. Instead, he will get angry and fixated on the origin and nationality of the employee."* With a non-response from the proprietor, her friend felt compelled to continue to rationalize the request. She went on, "We (the family) are just trying to make the remaining years of his life as bearable as possible. You are my good friend, so I can open up to you about this. He has been this way since I was old enough to understand his inclinations. He cannot change, sorry." I would not overanalyze this episode except observing that the ailing older man is spared some agony while he could wantonly inflict pain on another human. It happens!

Stretching this saga a bit further, here was a sick individual needing assistance yet could not temper his intolerance for non-native-born professionals. Does one's nation of origin diminish expertise, even upon receiving training in the United States? Would a sore muscle discriminate on the hand that massages it for wellness? Would medication administered by a newcomer work differently inside the patient's body? Of course not, but these attitudes sometimes occur.

You are simply exaggerating a couple of incidents, one could surmise. Well, let's continue. Kevin was an accomplished professional engineer who migrated from Africa to take a job with an aerospace company in the Northwest. He was highly recruited for his brainpower and innovative thinking. On the weekend, he went to purchase a relatively expensive instrument at a music store with his family in tow. Upon selecting the desired item, he showed it to the store manager of a different ethnic background. Without inquiring about the payment method, the manager immediately proceeded to shove the lease and multiple payment agreements for him to sign. Kevin was well-to-do and loaded, as they say, and would pay for thousands of such instruments with cash (check, of course, or debit card) instantly. The manager had assumed that Kevin could not possibly have had enough funds on hand to pay for the instrument. Therefore, he had prepared a leasing document that required a monthly payment plan over two years. You wonder how and why the manager reached such a conclusion. Kevin wrote a check for the purchase and walked away. Okay, I am still on my cloud, one could say. How about this one?

Manuel from South America, who resides with his family in Southern California, told a story of how, in his youth, he and his friends enjoyed themselves going from store to store playing some mind games. Sadly, a reflection of our society. Three of them would go into a store and immediately observe how the security personnel would quickly scamper and take positions, sometimes trailing them as they entered. He jokingly compared that to seemingly responding to a military base breach where all the sirens are blasting and all personnel reacting to an attack. One particularly hilarious experience he relayed saw two store security guards bumped into themselves as they were fixated on watching their every move in the store. He strongly felt that their ethnicity played a significant role in this supposed profiling scheme. Manuel and his friends were mischievous in their antics. They recounted when they would borrow Manuel's father's Mercedes sedan and drove around town to attract police attention. According to him, it was like magic. They were like magnets of the police. Once they drove a short distance, a police car would trail them and "manufacture" a reason to stop and pull them over. Sometimes, an amateurish officer would insensitively opine that they looked suspicious in such an expensive vehicle. They would always respond to officers' questions politely to avoid escalating the encounters. Once stopped, their driver's licenses would be requested and vetted by the officers. As naïve and perhaps silly teenagers, Manuel and his friends reveled in such pranks. Societally, though, it reflected a larger systemic problem of racial profiling.

Let us discuss a recent social profiling occurrence that captured the media's attention. Amanda Gorman, the U.S. youth poet laureate, who recited her poetry at President Biden's inauguration, alleged that a security guard followed her home and told her she appeared "suspicious." There may not be any other way to represent this other than profiling. Amanda is African American and young- two hydra-headed blinkers. Though female, it did not matter. The two earlier factors provided the warning signals to the guard. She protested, showed her keys, and buzzed herself into her building. The security guard tendered no apologies. One cannot advance a reason for the insensitivity, except perhaps the guard felt it was no big deal. Flat tire! Sometimes, intolerant behaviors can extend beyond generations.

How about a Sunday story? The most sacred day in the Christian week. Jonathan, who was highly religious even in his native European country, would never miss a Sunday worship. For context, Jonathan's original desire was to be a priest back home, but events and circumstances derailed that plan for now, as he would quickly add. Jonathan had missed Sunday service the first two weeks of his arrival, and that gnawed at him tremendously. On the third Sunday, he ventured into a church about two blocks from where he lived. He had no car, so he settled for the nearest church to worship his Lord, believing that it should not matter where one worships since we all worship the same God. Welcome to the reality of your new land! As Jonathan would recount the experience, he knew something was amiss when one of the church ushers accosted him at the door and wondered aloud to him if he had been lost. Remember, this was a church where the creed is that all are welcome. When he told the usher that no, he was coming to worship, the usher could not suppress his bewilderment and reluctantly showed him a seat at the rear. You see, Jonathan was Caucasian entering a black church to worship. He told the story that he was aware some segregation and even racism exists in his new land, but not at the church full of Christians, he had reasoned. What Jonathan had not fully realized and understood was that, as the late Martin Luther King Jr had once stated, "11 am on Sundays represented, and still represent, I may add, the most segregated hour in America."

Generally, we worship with our kind, ethnically and racially. That day's experience scarred Jonathan that he remembers it vividly even after thirty-two years. He could not wrap his head around how people that read from the good book, the Bible, and vow to live by its teachings violate the basic essence of belonging in the household of God. He lamented aloud how Christians could tolerate the divisions based solely on race and ethnicity. At the point of this discourse with Jonathan over dinner at a restaurant, I decided to rile him more by asking him to ponder the following. When he determined the answer, we would have dinner again, this time totally on me. Jonathan, in my opinion, is brilliant and curious. The type of mind and fortitude that pursues an answer to any nagging question. "Jonathan," I started, "How could Christians, especially Christian leaders such as pastors, priests, and bishops, practice slavery through the ownership of slaves? How could institutionalized religious organizations buy, deploy, and sometimes sell slaves? Did not the good book tell us that we are all God's children equal in His eyes and the world He created, with no one subjugating one another, except for the non-human creatures of the earth?" In case you thought for a minute that I made this up, read Galatians, 3:28, Genesis, 1:26-28, and Proverbs, 22:2.

Jonathan, I observed, had suddenly stopped enjoying his meal, as his brain and perhaps emotions were getting worked up, so I quickly changed the subject to a lighter matter, and we finished our meal and left. I avoided sharing with Jonathan that a new Baylor University study found that churches that attempt to create more racially and ethnically diverse congregations ultimately lose membership. The study concluded that churches that start as multiracial sustain growth than those that seek to integrate later on. It could be a good research topic for budding behavioral scientists to explore the "invasion" of churches by racially or ethnically different individuals. Such an examination would likely yield some interesting and rich outcomes. During these initial experiences for newcomers such as Jonathan, the air in their car tires is let out gradually. The culmination of these episodes results in a flat tire. With a flat tire, the car is not rendered useless or hopeless. Instead, it represents temporary immobility that can be fixed by replacing the tire. It does not permanently incapacitate the driver. In most cases, it helps the driver learn to deal with adversity and challenges. The hand they are dealt.

Anita, a vivacious, intelligent, and hardworking young lady, was promoted and positioned to take on the next level opportunity. Anita works for a national company that installs its manager and systems in client locations. She was offered a position in Pennsylvania that required her to relocate from the Southwestern United States. As was always the procedure, she flew to meet with the new client for pre-engagement familiarization and initial relationship building. After her trip, the client demanded that Anita not be deployed at their site. According to the client, Anita was highly qualified, had a very positive and infectious personality, "Her race and ethnicity would not fit well within our environment here," the client reasoned. I will not bore you with the resolution details, but Anita did not relocate to Pennsylvania. Again, Anita was well qualified for the job and had a can-do attitude to boot yet could not secure the position because of her race and ethnicity.

Remember that these subtle and sometimes blatant undercurrents are always present. The newcomer is now coming to grips with them and baptized with America's realities and paradox. How could a society so encouraging and supportive of liberties, self-responsibility, accountability, hard work, etc., be equally stifling, oppressive, and sabotaging. Here lies the contradiction of America. Allow me to quickly state that no society on earth is bereft of what I will loosely refer to as the dark side. All nations do, some lightly, some massively. But on balance, most, if not all immigrants would endure the dark sides of America. After all, they recognized these underpinnings, even from afar, before embarking on migrating to their new land.

Flat tires could come in different fashions. We have identified the pronounced racial and ethnic tensions, behaviors, and attitudes. There is a corollary to the racial and ethnic struggles. It manifests in the playing field not plainly leveled for all. However, it could be argued and recognized that tremendous strides have been made in this area. However, more work remains. Consider the lending sector. If you have a trusted friend, family member, or acquaintance, seek to be honestly told the dark side of lending, especially as it pertains to ethnic minorities and immigrants. Sometimes, it is cloaked as a lack of credit history, etc. How could such a history be built outside of the United States? Understandably, some filters must be required and applied by lending organizations in assessing risk. However, some lenders utilize this barometer arbitrarily to deny credit and funds to well-meaning and qualified individuals. I must say that some industry and government stipulations have streamlined and brought more objectivity to the lending process, I am told.

To the academic environment, we go. Benedict, popularly known as Ben, had emigrated from Africa. Ben was a brilliant and studious student. While in Africa, he excelled in his studies posting the highest grades in almost all subjects. His command of written and spoken English was superb. Due to his grades, manifested in his transcript, he was admitted to study Accounting and Finance at an American University in the Southwest USA. He was very excited and reported to the college. As was the school procedure and protocol, he was assigned a counselor cum adviser to shepherd him through academic decisions. The office receptionist directed Ben to a building to meet his adviser. Ben successfully located his adviser's office and waited outside her office until she became available to meet with him. When the adviser was ready, her assistant beckoned Ben to enter the adviser's office. Once Ben was seated and after the initial courtesy greetings, they went down to business.

The adviser opened Ben's file and briefly perused the contents. She asked Ben a few questions about his academic and ultimate career goals. Ben proudly and confidently conveyed his ambitions and aspirations. They then considered the initial classes that Ben should take. The adviser slowly began by telling Ben that his prior academic records were impressive, including his English language grades, but there is a problem. Ben's heart started to sink. She continued, "Quite honestly, I was straining to understand you when you speak. Therefore, you will have to take remedial English before enrolling in College English." Back home, Ben was very proficient in the English language, Oxford English, mind you, which is held as superior English. Yes, Ben spoke the English language tinged with his accent, which sometimes took focus to comprehend fully. Ben never had a problem with the English language. His expression was adulterated by his accent that got in the way. So, Ben got offended by the put down and quietly protested by arguing that he was very fluent in both written and oral English. His non-acceptance of the adviser's insinuation and directive fell on deaf ears. Oh no, flat tire!

Ben was crushed. The adviser's directive meant that Ben had to spend more money paying for non-college credits and cause a delay in his potential graduation timeline. Ben had always been a fighter. He decided he had to go to battle. Ben strongly believed that due to the adviser's seeming struggle in correctly deciphering his accent, she set him back unfairly. He agonized over what he termed an arbitrary and subjective judgment of the adviser rooted in her lack of adequate cross-cultural exposure and competency. Two days later, upon the advice of an American-born potential classmate, whom he had shared his predicament with, Ben wrote an impassionate and elegant letter to the Head of Students Affairs at the University. In the letter, Ben proposed that he be allowed to enroll in the College English class, and if he were failing after one month, he would withdraw and revert to the non-credit preparatory English course. Should he kept pace with the College English class, he would be allowed to continue on that path. After reviewing his letter and in consultation with his adviser, the Head of Students Affairs approved Ben's proposal. Ben not only completed the College English class, but he also came out in flying colors with an A- grade.

What occurred here? You may wonder. There was a convergence of factors both from a newcomer and entrenched citizen perspectives. Ben was very proficient in both written and oral English but possessed an accent that he could not easily shed. The adviser lacked cross-cultural understanding and impatiently assigned Ben into a category he did not belong to. She would not dig beyond the surface in search of the gem. It was easier to cast and move on, even if improperly. What was disheartening about this ordeal was that the adviser was supposed to have been his advocate, Ben bemoaned. Instead, she was just another college employee who had not applied her trade painstakingly. Newcomers are sometimes subjected to these experiences intentionally or inadvertently.

Some emigrated to the United States with their families, including wives, husbands, and children. Children have been known to adapt to their new realities quicker. The external exposure to the school becomes the fastest accelerator of their assimilation into their new society. The children's speedier path to acculturation creates tremendous tension and stress for their parents. Before the parents can bat an eye, the children's language and vocabulary have shifted to mirror their school friends' and classmates' lingo. They absorb the pop culture of their new land at a dizzying pace. The chasm between them and their parents starts to emerge. On their part, their parents continue to cling to the old country's ways. The children exacerbate the situation by labeling their parents archaic and out of step. They are ignored and irrelevant and no longer the anchor for the children like they were before emigrating. Oh no, we have a flat tire! The described dynamic is more prevalent than can be imagined. So much so that it generally tears at the fabric of the family unit. The parents usually fight to maintain the so-called family cultural core, but it is a lost battle. The children have now morphed into pseudo-Americans, while their parents remain in this wonderland of wishes – more on this in chapter 7.

Most immigrants enjoy a more extensive family system than natural-born Americans, generally speaking. It was this network of relations that sustained them in the old country. When a doctor's appointment is scheduled, a cousin, nephew, or distant relation would come and watch the children on an afterschool schedule. During business travel, a family member is brought in to babysit the kids. Should someone become ill and needed assistance, the family quickly summoned an extended member to nurse the ailing individual. All of those kinds of support evaporate in the new land. Indeed, such a reliable subsistence system is forever gone, vanished, stamped out. You are now left to fend for yourself and your nuclear family. Oh no, flat tire!

The so-called flat tires know no boundaries nor class. Consider the following that was widely reported by the media in October 2020. The treatment meted to the second lady of the State of Pennsylvania. The wife of Pennsylvania's lieutenant governor was openly insulted with a racial slur while grocery shopping due to her newcomer status. Gisele Barreto Fetterman, married to the State's Lt. Gov. John Fetterman, was born in Brazil and had emigrated to the United States. The verbal assault on her was open and brazen, she recalled. She reported to authorities that a white woman accosted her in the store. As she stood in the cashier's line, the woman lowered her mask (it was during the COVID-19 pandemic period) and uttered, "Ugh, there's that n-word that Fetterman married. You don't belong here. Go back to where you came from." What made this incident troubling was that the insulter came back two other times to haul more intolerant words at Mrs. Fetterman, additionally at the line, and finally in the store's parking lot as Mrs. Fetterman was driving away.

Furthermore, the woman knew who Mrs. Fetterman was and her background as a newcomer to the United States. It did not matter to her that she was related to the State's Lieutenant governor. Oh no, flat tire! The woman's tirade was recorded this time at the parking lot, even as the intolerance-spewing woman observed the cellphone video recording with some glee. The State's second lady had casually gone to the store without her security detail. Misunderstanding, ignorance, and intolerance abound.

You will be grossly mistaken if you think that racism and ethnicism are the prerogatives of the Caucasian race. Not even close. Other races in America and worldwide also perpetuate this intolerant and abhorrent behavior. Remember the ethnic cleansing in Rwanda, Africa, or the ethnic genocide in Bosnia and Herzegovina, Europe? Humans, regardless of their location, possess an uncanny ability to mistreat and discriminate against other humans. Any group of people with some societal advantage usually suppress other groups they feel are inferior to them. Behaviorists attribute such behaviors to a sense of insecurity or the need to maintain power and control. Others have blamed scapegoating and frustration for these negative behaviors. Whatever the reason, it exists.

Let us return home. Apart from institutions such as churches, other establishments abound that are loosely governed by race and ethnicity. Ever visited a bar or nightclub reserved for African Americans? A Caucasian entering one of these spots is starred stiff. They go at their peril. A college friend told the story of his observation at a popular "black nightclub" in Chicago. While standing in line to be processed for entry, he heard one of the bouncers at the doorway admonish a Caucasian pair to reconsider entering. He said, "Dudes, are you sure you want to waste your money paying to go in? You will not enjoy it. You don't belong here." Ouch! The bouncer's basis for his assertion was race and ethnicity. Reaching such a conclusion strictly on race was unjustifiable. It happens. By the way, should you be interested in gaining some race education without paying any tuition, visit your local black barbershop. This platform represents perhaps the most sociological forum for light and heavy subjects in the Black American subculture.

Let us talk about an item that is taken for granted but sometimes emerges as a mental health issue. Unless one emigrates from a similar geographical latitude, chances are the weather would be different from that of America. Yes, even in America, climate and weather differ significantly among regions. However, imagine you are emigrating from a tropical, temperate, arid, or mediterranean climate, how different and perhaps traumatic the weather in your new land becomes. Some studies have linked certain mental health syndrome to depression brought about by weather impact. Not only would the newcomer learn how to adequately dress for weather conditions in his or her new land, but he or she would also navigate traveling under certain weather conditions. Flat tire!

This one could be classified as a mind-bender. American economic and societal systems are generally envied the world over. The apparatus and infrastructure of the societal engine are solid and fully functional. However, Juliette, who had emigrated from France about eighteen years ago, saw it differently. According to Juliette, her initial observations and feelings when she migrated to the United States were disappointment and frustration. She complained about how things seemed inefficient and clumsy. Upon further probing, she advanced some examples to bolster her conclusions. The one area that heightened her negative feelings was securing an apartment in New York City. She explained that it was more arduous to navigate obtaining a place to live in the U.S. than her old country or for that matter in Britain, she claimed. She had lived some years in Britain when she was a young adult. She stated that an apartment rental process in the United States was more intrusive and cumbersome for the potential tenant. Ranging from the volume and depth of personal information sought to the degree of future earnings. To her, other similar urban and cosmopolitan centers around the globe demand lesser information and data for access to a place of abode. Juliette's New York City experience may not necessarily be similar to a good old mid-western USA ordeal. Juliette further noted that the United States Postal Service operation seemed like that of a third-world country. She intimated that her initial experiences reflected a sluggish and unproductive postal system. Lines were usually long and slow-moving. Oh no, flat tire!

Juliette was queried about her current feeling, having lived in the United States for eighteen years. She commented that her initial experiences indelibly imprinted negative connotations and associations that she could not shake. She admitted the intensity of her negative feelings had waned over time. Her overarching conclusion was that money ultimately provides access to things here in the U.S. for the privileged, and the rest of the population just tries to survive. I had referred to Juliette's take at the beginning of her story as a mind-bender, and let me finally explain why I termed it so. You see, all things considered, America and her society tend to represent an open society where greater access and opportunity are more readily available for its citizenry. Yes, there are inequalities within the system, but America is regarded as a nation possessing general access to basic items, comparatively speaking. For example, products and services at stores are presented for whoever can acquire them, rather than restriction only to the well-heeled. Of course, one could argue that the lack of economic power is a restriction in itself. My point here is that Juliette's contention was geared more towards the blatant and open class and access demarcation generally found in the so-called third-world countries. From that standpoint, America does not belong in that category, I submit.

Flat tires are not permanently debilitating but must be addressed for the vehicle to continue its journey. The speed and efficiency of fixing the tire are tantamount to the traveler's tenacity and focus. For the newcomer, recognizing, assessing, and resolving these inherent challenges determines how successful your journey becomes.

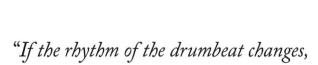

"If the rhythm of the drumbeat changes,
the dance steps must adapt."
-African Proverb

CHAPTER 6

MUST ADJUST MIRROR

(Readjusting the Initial Adjustment)

As a new entrant, you face these seeming challenges and obstacles. Do you fold and hide, or do you come swinging positively, of course? At this stage, one recalibrates expectations, readjusts mentally, and repositions spiritually. Some crumble in the crucible of the new environment's pressures, while others grind through it and emerge unscathed.

We will examine some scenarios experienced and lived out by new entrants. Let us call the first migrator, Sam. Sam emigrated to the United States full of promises and great desires. Sam was also capable, both intellectually and physically. His family from his old country was middle class and fairly educated. So, came Sam into his new country. He started with all the anxieties of a newcomer. Soon, Sam found his way around and assimilated at a faster rate. This rapid integration was because Sam was a social butterfly. He was amiable, with no hard edge, and remarkably adaptive. One could argue that his more effortless flexibility became his undoing.

Life in Sam's new home was taking shape. Then came all the forces discussed in the earlier chapters of this book:

land of incredible opportunity and the proliferation of good and bad life elements, including illegal drugs, racism, and intolerance. The unsurpassed liberty and freedom and the trappings of a cosmopolitan environment, all came calling. Sam lived in New York City. The confluence of all these tensions fell Sam. Sam was introduced to recreational drug use by some coworkers at his part-time job. He dabbled into light drug use and soon elevated to using harder drugs. The wheels had begun to come off. Sam crossed schooling off his priorities and soon, legal and routine work became too tasking and laborious. Easier life was beckoning, resulting in fraudulent and criminal ways of attaining such a lifestyle. Sam had now morphed into a medium-sized neighborhood drug dealer, arranging shipments through drug moles and distributing to the smaller dealers. Ostentatious living became the norm for Sam. His very plush penthouse became the weekend party place for his friends and "business partners." Life seemed good, in fact, superficially great for Sam. In the meantime, Sam's relatives at his country of origin believed Sam was earning his academic degree in his new land.

One September early morning, at 4:15 am to be precise, five federal Drug Enforcement Agency (DEA) officers surrounded Sam's building and knocked at his door. Sam was sleeping with one of his many girlfriends who often hung out with the big-spending Sam. Sam was whisked away under arrest in handcuffs. The DEA had been monitoring Sam, taping his phone, trailing him, and watching his every move for the past eight months, even including a couple of international trips that Sam took. The Drug Enforcement Agency (DEA) later revealed all relevant connections, networks, and sources that Sam was utilizing via surveillance. Therefore, when they swooped down to arrest Sam, the case against him was airtight. Sam was subsequently convicted and served numerous years in the federal prison for illegal drug dealing. Sam never recovered. Though released after serving his sentence, he drifted and is much older and in poor health.

Sam's odyssey illustrated a newcomer who succumbed to his new society and environment's pressures and lures. No one knows if Sam would have derailed from his erstwhile stable and good-citizenry-focused life he had lived before straying. Social scientists would have a field day debating if the environment made the man, or the man was already inherently made, or a combination of the two. We leave such discourse to the researchers but not without noting that scholar L. Todd Rose submitted "Behavior is not something one has. Rather, it emerges from the interaction of a person's biology, past experience, and immediate context." One thing was clear with Sam's predicament, though, and that is that the conglomeration of the forces of change, new exposures, new experiences, and conflicts of two cultures (his old country and the new home) exacerbated Sam's situation and, therefore, accelerated his downfall. Sam is not alone in this struggle. A large percentage of newcomers face these ugly truths and challenges. They come in different flavors and from diverse angles. Let us explore the saga of Elizabeth, popularly known as Lizzy.

Lizzy emigrated to the United States at age twenty-nine. She had struggled to eke out an economically respectable life in her country of birth. She applied through the appropriate visa program to legally migrate to America. Lizzy is an extraordinarily talented software engineer. Not only did she graduate summa cum laude with her bachelor's degree, but she also earned a perfect 4.0-grade score for her master's degree. She was highly recruited by Silicon Valley, which was where she ended up. Lizzy was effusive about her migration and was full of life. Gregarious and friendly, she fit quite well in her new environment.

Then came the romantic and love bug. She soon fell in love with an associate in her town's computer world. They had started as casual business acquaintances, then developed into good friends, which led to a budding and intimate relationship. Things seemed normal and natural. Lizzy had now been in her new adopted country for about two years. She had bought a small cottage and settled in. Living by herself meant she relied on a handyman for minor repairs at her home. This necessity led to an area handyman referred to her by a coworker. The handyman performed all tasks and repairs diligently and at reasonable charges. However, Lizzy did not know that this handyman was an ex-convict who seemed to be turning his life around. He was convicted of bludgeoning the then-girlfriend to death in another state. After serving his time in that state, he relocated to California. The coworker who referred Lizzy to the handyman was not aware of his violent history. He had been referred to the coworker by an acquaintance at the local gymnasium some years past.

Lizzy living alone and being an attractive lady was too much for the handyman to ignore or resist. The handyman's history, coupled with the lure of a vulnerable single lady, had triggered evil thoughts in him. It was an autumn night with the bay area's cool breeze blowing lazily. Lizzy had prayed and tucked herself in for the night and must have dozed off, she recounted. Suddenly she was blinded by the flash of a powerful touch light and could not decipher who the holder was. As she started to scream, the assailant muffled her with some thick fabric. Her mouth and eyes were immediately duct-taped, and she was sexually assaulted. It did not end at that. The assailant utilized the thick cloth to try to asphyxiate her. Lizzy was intelligent and clever. At one point during the ordeal, she faked dead and motionless. Sensing that she was dead, the assailant fled. Without hearing any sound or movement after about ten minutes, Lizzy crawled out quietly from her bed and called 911. She was subsequently rushed to the hospital for treatment and observation. She later returned home after two days of hospital stay. Lizzy's life had been spared but changed forever. The ordeal was so traumatic that she took some time off from work to recuperate.

While the police were investigating the crime which had led to the handyman, who was now under arrest, Lizzy received devastating news. The rapist had transmitted HIV to Lizzy, who tested positive for the virus after about seven months since the rape. As a suspect with incriminating evidence, the handyman was tested, and the results came out positive for the virus. Bear in mind that this was in the eighties when HIV and associated stigma was at their peak. As if the rape trauma was not enough, Lizzy was dealt another blow with the HIV infection. Life seemed hopeless for Lizzy at that time, and she never fully recovered from this dual knock on her life.

No one knows if this could have happened to Lizzy in her old country. We cannot second guess providence. Did Lizzy's background and culture of trusting and her prior experience with lower crime rate lead her into complacency with her safety and security, especially while living alone? Could she have more thoroughly researched the handyman's background, now the assailant who was eventually convicted of the crime? Sometimes, the upbringing, culture, and society of a newcomer blur the dangers of his or her new community. Sometimes, before the new entrant had fully absorbed, understood, and appreciated the good and dark sides of one's new environment, it becomes too late. Bad and terrible things happen. The newcomer's naivete and innocence make one vulnerable to the harmful elements of the new society. This scenario of the ineptitude of understanding one's new environment can be costly and sometimes deadly to the newcomer.

How about this vehicle insurance story? Suki emigrated from East Asia. According to him, obtaining and maintaining vehicle insurance is not strictly practiced and enforced in his old country. Usually, drivers bribe their ways when caught violating the insurance requirements. What is more, insurance companies typically wiggle out of fulfilling their financial obligations when needed. Drivers, therefore, are very cynical and untrusting of vehicle insurance programs. Suki grew up with this distrust of insurance and related matters. He carried this attitude to his new land, America.

When Suki bought his car upon arrival in the United States, he quickly learned that one must obtain insurance before driving out with a purchased vehicle. He did just that. However, when it came time to renew his vehicle insurance, Suki conveniently ignored it. He was driving around without insurance on his vehicle and liability for external damage to anyone or their vehicle. His luck ran out. One rainy and foggy midmorning, he rear-ended a car and caused substantial damage to both cars. Besides that, both injured drivers were taken by ambulance to the hospital. Upon release from the hospital, Suki was engulfed in a protracted legal and financial liability that drained his finances for many years. Suki's background influenced his decision in his new land not to maintain his vehicle insurance. Newcomers must learn their new land's uncompromising culture, norms, and practices and quickly adapt to them. Suki learned the hard way. He has become a strong advocate for quick adaptation to your new land's norms and practices. He does not hesitate to counsel newcomers to adhere to all rules and regulations regardless of their beliefs and experiences in their old country.

Some newcomers arrive in their new land excited, straitlaced, and bubbling with life. They embrace and plunge into their new land with all vigor and aspirations. They record successes and make substantive inroads in pursuing their American dream. Due to certain reasons such as lack of steadfastness, focus, and sheer laziness, they succumb to the deceivingly construed easier path of utilizing and "milking" the governmental support systems. The system was established as a safety net for the downtrodden and financially struggling population. They quickly learn how to bilk the system without exerting personal efforts and hard work. They tap into the welfare pipeline and access the disability programs in some cases. This behavior's sad commentary lies in the sapping of the self-reliant and individual doggedness spirit. The enticement of the societal safeguard for the truly needy becomes too tempting to resist. They must exploit them. The other pitiful observation with this scheming is that it requires elaborate planning, lies, and deceit to pull it off continuously. Some newcomers who get entrapped or consumed by this mendacious approach towards pursuing their dreams make course corrections and refocus their lives. Others are disciplined by law for their transgressions. They chose to deviate from the legal hand they were dealt. It happens!

Let us pivot to some good outcome stories. Anthony had earned a doctorate in Nuclear Physics in his country of birth before emigrating to the United States. The U.S.

Atomic Nuclear Agency recruited him. He legally moved to this country with his wife and two small children. Anthony was laser-focused to succeed and thrive in his new land. He obeyed all laws and availed himself of all resources designed for resettling new immigrants. Anthony was not sparred of the tensions, conflicts, and dichotomous situations littered in the new entrant's pathway. Yet, he chose the good route towards actualizing his dream of earning a decent remuneration for his work and attaining the so-called American dream. Anthony achieved tremendous success and recognition by all intents and purposes because he kept his nose clean and failed to succumb to the hidden yet powerful pressures that faced every newcomer. He was intentional, disciplined, and determined to capture the golden fleece. His wife enrolled in college, obtained a degree in Psychology, and taught at a charter school in their community. Their children were raised with a Christian framework and grew to become solid and civic-minded citizens. The family remained happy and an integral part of their local community and church. At one point, Anthony contemplated running for a political office to represent their district in the State Senate. At different times, his wife held positions in the local library board, the Parent Teacher's Association, and their subdivision homeowners association. They successfully represented and epitomized their community positively.

Let us review one more scenario. As an exemplary student, Festus emigrated to this country to continue his education. He took advantage of all financial aid and loan programs readily available to anyone who desired to further their education. Festus was promoted several times at his job with the educational successes achieved. Besides, he assimilated into his community somewhat seamlessly. He was a member of the local country club, where he played and enjoyed his golf outings and games. What had really occurred to Festus, Anthony, and their likes? Why did they not derail under the pressures and tensions of a new society and the added initial cultural conflict? How did they navigate the landmines in their new land? The answers to these questions are not far-fetched. At the core is how one grapples and handles the hand one is dealt. Anthony and Festus's stories amplify and represent great testimonies of newcomers successfully harnessing all available tools, programs, and opportunities at their disposal. The hand that is forced on the new entrant is often unintentionally wrought with embedded societal pressures. These elements are compounded by the sociological, cultural, and sometimes religious tugs on the newcomer. The newcomer cannot tenaciously cling to his or her old country's cultures. At the same time, we are constituted by our background, which includes cultural underpinnings. These foundations are not easily discarded or peeled off. The struggle is real and severe. Imagine if you were born and raised in, say Iowa or Indiana or for that matter, any part of the United States, and you emigrate, to say, Africa. You could not easily shed your American values, culture, and mindset, even if you tried. This clinginess is because your background had already formed who you are. Yes, there could be some successful adjustments made with your grounding, but they usually are at the periphery. Social scientists tell us that our personality and character are fully developed by age seven or thereabouts. And that our surroundings and some inert traits blend to pronounce who we are or become. The famed futurist and philosopher Alvin Toffler submitted that "the illiterate of the 21st century will not be those who cannot read and write, but those who cannot learn, unlearn, and relearn."

So, it is generally challenging for humans to unlearn and relearn things. Unfortunately, such a vulnerability manifests when we migrate to a new land. Failure to adjust greatly impacts assimilation and acculturation needed to succeed and thrive. It is often said, "When you are in Rome, you behave like the Romans." This adage is easier said than done. Try uprooting, discarding, or outright suppressing your being's fundamental pillars and tenets and see if it is an easy undertaking. It is not impossible but remains herculean. The point here is that altering our worldview must be done with care and intentionality for a positive and wholesome outcome. While the newcomer is grappling with this tussle, the new land's societal pressures do not abate. Indeed, they magnify and become impossible to ignore. Eventually, the new entrant navigates this self-churning ordeal. The initial adjustment is modified.

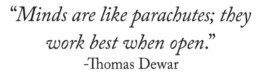

"Minds are like parachutes; they work best when open."
-Thomas Dewar

CHAPTER 7

DIG DOWN INTO
THE WELL!

(Deployment of Coping and Thriving Skills.)

The newcomer is now at a crossroad. You must do something to continue to pursue your dream of success in your new adopted country. You must not biblically pour old wine into the new bottle as found in the following bible passages: Matthew 9:17, Mark 2:22, and Luke 5:37. It is a new day and a new awakening. You always knew this, but now the rubber meets the road. So, what do you do? Well, you reach into your arsenal and dig down to scoop up coping and thriving skills.

Let us start with language. You assess your language skill level vis-à-vis your academic or career path. You begin to work deliberately to bridge any gap. Language skills do not just entail speaking English fluently. It encompasses immersing in verbal and non-verbal language competencies. You become conversant with slang and local vernacular. You understand and appreciate expressions such as "chill or chill out," meaning calm down, "to bail," referring to leaving, "couch potato," indicating someone who spends little or no time exercising and spends a great deal of time watching TV. "Down to earth" means practical, genuine, and realistic, "cold shoulder" refs to intentional unfriendliness, "plead the fifth" entails not incriminating oneself, "screw up" means mismanage or mishandle a situation, and "a creep" is for someone lacking respect for personal boundaries. While we are on it, let us explore some more words, phrases, and sayings. "Piece of cake" means super easy, "John Hancock" is an American synonym for signature, "taking a rain check" means rescheduling or postponing an appointment for a mutually agreeable date. "Shoot the breeze" refers to engaging in idle, empty chatter, "throwing one under the bus" means betrayal for one's self-interest and advancement, "don't be such a wet blanket" means do not be a downer or ruining everyone else's good time. "The cat's out of the bag" or "spilling the beans" both refer to revealing a secret. "For the birds" means trivial or worthless, "don't cry over spilled milk" means upset over something you cannot fix. "On point" for something done well and perfectly, "slay" means mastery and best of the best, and finally, "touching base" means getting in touch with someone. These are common and everyday phrases that aid incursions and a fuller understanding of your new land's language and culture. Strong connections have been made for centuries between culture and language. One could posit that they are inextricably intertwined. One feeds unto the other and vice versa. So, by learning some peculiar language of your new land, you are invariably enhancing and accelerating your assimilation. Mariam's experience with slang, in chapter 15, illustrates the urgency for immediate familiarization with the local language and subtleties.

Depending on which country one emigrates from, the chances are that many of these countries' economies are not modeled on credit and associated components. Some of these countries utilize cash and liquidity as their economic engines.

EMEKA OKEANI

We will reserve a discourse on such a system's advantages and disadvantages for another time. The point here is to highlight that the newcomer had been used to an economic system built on cash and liquidity. You delay purchases in such a system until you save enough for an outright buy. You defer gratification from goods and services until you can instantly acquire them. Lo, you are now in an economic system where you can obtain your desires while not wholly paying for them. Of course, they are not free, but you are not subjected to suppressing your wants and needs because you do not possess the total funds to secure them outright. Again, we will not engage in discussing the system's merits and pitfalls. The economists and financial gurus can debate that to their heart's content.

You are now in an economic system with the latitude, means, and opportunity to fast-forward the acquisition of goods and services and pay later and gradually. As a result, you must learn the art of regulating your earnings to allow for the periodic, usually monthly, payment of your bills. You must shed the hitherto associated guilt of enjoying partially paid-for goods and services. Your new land's economic system does not associate any stigma or shame towards such practice. It is the norm and an integral part of propelling the economic engine. Indeed, you are rewarded for partaking in such an "enjoy-now-pay-later dynamic." Some new entrants relayed how it took them several years to become used to this model. They initially argued and contended that the system forced them to live above their means which they fought vigorously, both psychologically and philosophically. They succumbed to this established and flourishing system after realizing that to survive and thrive in their new land, they had to play the hand they had been dealt. Thus, began the process and path of building their creditworthiness to access the economic opportunities that existed and made the United States the envy of the world due to its economic vibrancy.

Unless you emigrate from some of the violent hot spots of the world rife with wars, unrests, and strife, you may not be used to or exposed to guns. Due to the freedoms and liberties fought for and enjoyed by its populace in your new land, guns are everywhere. At Walmart and area department stores, numerous guns are on shelves and drawers to be bought. What is more, all you need to own one is a government-issued identification, proof of being a law-abiding citizen of legal age, and you can walk away with one. Your new and adopted country cherishes and revers the 2nd amendment law and the right to bear arms. With this culture and way of life, anyone can use their weapon anytime. Your antennae for safety and security are heightened. You become more safety and security conscious than you had ever been in your life. You even start to contemplate buying a gun for your family's protection. You purchase one for the primary psychological reason and perhaps for protection one day.

Florence, who had emigrated from one of the Nordic countries, told of a horrifying experience on the streets of Houston, Texas. According to Florence, she witnessed a road rage, which resulted in an afternoon shootout by two drivers, with one dead after the confrontation. She was driving beside one of the combatants with her two-year-old strapped in his car seat. She saw everything. The incident so shook her that she occasionally had nightmares. She stated that although she and the family had lived in the United States for eighteen months before the incident, this occurrence quickly made her realize that she now lives in a new land. She claimed that she had never seen a gun near her life, what more seeing and hearing guns discharge up-close. And not just one but two guns and shooters. As a newcomer, you must recognize your new land culture, which includes the gun culture. America's gun culture encompasses beliefs, attitudes, and predispositions towards firearms and their accessibility and use by the populace. This culture is rooted in hunting, militia, and a frontier lifestyle. The American Revolutionary War and its success buttressed this embrace and love of firearms and their proliferation. Some, though, submit that gun culture was planted long before the Revolutionary War. They maintain guns, and all firearms were amply deployed as the initial European settlers made their incursions into North America. Since then, gun culture has waxed strong and remains endearing. American exceptionalism has many unique components, and the rugged individualism which the gun-toting culture symbolizes endures as one of its strong anchors.

American economic system is built on capitalism. One of capitalism's core tenets is its reward component for work and effort. One of the subsets of work and effort is personal responsibility and accountability. This mentality and practice have powered American ingenuity, innovation, and progress for centuries. All one has to do is look around. You will observe how personal responsibility and accountability produced breakthroughs, such as those from Albert Einstein and his footprint on some scientific theories, Thomas Edison, Alexander Graham Bell, Henry Ford, Patricia Bath, Steve Jobs, and many more or business pioneers and achievers, such as John D. Rockefeller, Madam C.J. Walker, Andrew Carnegie, Walt Disney, Bill Gates, etc. The personal responsibility, accountability, and doggedness of these Americans launched them to these high heights.

Additionally, the personal responsibility and accountability bent of the United States culture produces an orderly society. Allow me to illustrate this. Venture traveling to some bustling capital cities, especially in Asia and Africa, and you will experience chaotic and hellish traffic. These cities are usually teeming in populations and busting at the seams in all measurable factors. Though there are operational traffic lights and sometimes law enforcement personnel around, these lights are hardly obeyed. The lights seem placed for decorative purposes, one would conclude. Motorists, motorcyclists, and pedestrians ignore these government-installed gadgets and apparatus meant for the citizen's safety and orderly flow of traffic. These installations were predicated on the citizens exercising personal responsibility and accountability in obeying the signs. Remember in chapter 2 when Cecilia marveled at the motorists meticulously obeying the traffic lights on her arrival day to the United States. If you recall, she was fascinated, surprised, and bewildered by such obedience to the laws and society's orderliness.

Let us explore another situation. In some migration countries, obediently and respectfully queuing in line for services is for the birds. Cutting in line is the norm, and the boldest or more aggressive push themselves to the front of the line. In the United States, as in most orderly societies, organized queuing etiquette is observed and practiced. So, for the new entrant contemplating bulldozing their way to the front of the line, a significant behavioral adjustment is imperative. You will be called out should you violate this basic rule of personal responsibility and accountability.

Diversity and inclusion are constructs and subjects that can be complex as they seem simplistic. This misinterpretation is because there are many tentacles and complexities to these tandem topics. America has often been referred to as a melting pot. This nomenclature stems from the amalgamation of its population's diverse ethnic and racial groupings; America's history and journey as a nation brought about this happenstance.

Interestingly, the United States is probably one of the world countries with the most diverse populace. This feature represents a blessing and also a curse. It is a blessing because diverse peoples produce a rich and thriving environment and a pitfall if the various groups cannot optimally harness their differences. Therefore, the chances are that the newcomer might have emigrated from a more homogenous country and community than the United States. So, the new entrant must cultivate and practice diversity and inclusion in the more diverse community they now belong to. Diversity and inclusion are often viewed and treated as the same concept. They are not. Organization theorists and practitioners consider diversity as the "what" and "who," and inclusion as the "how." The "what" and "who" refers to the variation in personal, physical, and social characteristics that make up the group, such as gender, race, ethnicity, age, sexual orientation, education, interests, etc., and who is sitting around the table. Inclusion represents the vehicle that enables diversity to blossom. It is the culture that allows all participants to engage and thrive. It is the behavior that welcomes and embraces diversity.

Furthermore, inclusion comprises integrating everyone and enabling the differences to coexist to elevate all members. It is a condition of feeling accepted and comfortable in the setting. Diversity is held as inviting differences, while inclusion offers equality by presenting a fair chance to all. An organization or group may be diverse but not inclusive. These groupings may possess diverse backgrounds and profiles, who may not fully participate in the stakes. So, whereas they may be mixed in composition, they have not enabled them to be fully participative members of the organization or society. Instilling and practicing inclusion requires intentionality and purpose. For example, during company parties, does the organization provide alternative food items for vegetarians or those who, for religious reasons, cannot consume the meats offered? The recognition of the group's diversity and providing a wide array of food items to cater to all members' needs and palates represent inclusion. A part of coping, succeeding, and thriving in the new society involves learning how to work, interact, and relate to neighbors, coworkers, and fellow parishioners in the new land. Some newcomers retreat to their cocoon by concluding it is easier than dealing with the hard truths of opening up their minds, perceptions, and worldview. It is stated and observed that the longer the newcomer shies away from engaging in their new community, the longer the assimilation and acculturation process and journey.

You may have heard the saying, "work hard and play hard." I will paraphrase it by re-stating: work hard and enjoy yourself. The American culture heralds hard work and rejuvenating oneself for even harder work. Therefore, self-renewal is at the heart of continuous effort and hard work. Citizens reward their hard work through vacations and retreats. Most cultures also cherish and practice self-decompression. It is an annual ritual of high importance to individuals and families. The newcomer must learn that they are expected to take some time annually to sharpen the saw proverbially. This regeneration is a part of the work-life balance of your new land. Noted author and self-help guru Stephen Covey amplified this notion of self-renewal in his award-winning book, *The Seven Habits of Highly Effective People*. Immigrants from most European countries have an easier time with this cultural underpinning. This rebirth's appreciation is because, more likely than not, their old country's culture may be parallel with this American workplace feature. Some other cultures may struggle with the concept of saving one's hard-earned money all year long only to blow it on a two or three-week junket, no matter how invigorating or self-adulating that might be.

Seeking and belonging to affinity groups in your workplace or community are highly encouraged. These groups comprise members who share common backgrounds, values, and possibly worldviews. Tapping into the experiences and advice of forebearers could be invaluable. As a newcomer, James was struggling tremendously in his personal and work life. Life was coming at him at jet speed, and he was quickly running out of fuel and suffocating. One day, as he talked on the phone with his cousin in a far-away city, he confessed to being in a quagmire with all the challenges he was facing as a newcomer. His cousin narrated how belonging to an affinity group in his community assisted him tremendously when he emigrated to the United States two decades earlier. James sought and obtained membership to a local affinity group with this piece of advice. As James would tell the story, he claimed that the affinity group did for him what no one, no relation, no organization did. In his words, "I found my compass through the affinity group." James became an unabashed proponent and cheerleader of affinity groups and encouraged newcomers to belong to one quickly through his experience.

It has been advanced that organizations unknowingly engage in unconscious bias by hiring for a potential candidate's cultural fitness. Candidates should be evaluated as culture-adds instead of culture-fits. While similarities of interests and backgrounds should matter, they should never represent the deciding lever. I bring this up here to admonish that organizations must be sensitive to avoid being blinded by the desire to load up on the employee's similarity rather than their differences that create a richer environment. Its quest to attract and retain individuals that could snugly fit an organization's culture could unwittingly create a monolithic environment.

Management thinker and author Rosabeth Moss Kanter postulated that we tend to attract and hire people whose characteristics, worldview, and attitudes mirror ours. She reached this behavioral conclusion in her book, *Men and Women of the Corporation*, written about four decades ago in examining gender inequalities within the corporate workplace. She advised we should challenge ourselves by thinking outside the building. Kanter emphasized that solving or tackling any significant change initiative rests in the power of ideas and working across sectors, industry, stakeholders, and interest groups. We must learn to become comfortable with seemingly uncomfortable situations and options. These choices may be different for our organizations and us. Therefore, leaders must be acutely aware of this danger and guard against it adequately by embracing diversity and inclusion of thoughts and perspectives.

*"Being challenged in life is inevitable,
being defeated is optional."*
-Roger Crawford

CHAPTER 8

HUMAN OBSTACLES: BLATANT AND SUBTLE IMPEDIMENTS

Humans are described as social beings in their evolution. We interact with each other under a complex network and nexus. In our homo sapiens relationships, we exhibit behaviors and tendencies that are inert and learned. These include good as well as unflattering proclivities. In this chapter, stories that speak to human failings relative to attitudes meted to some newcomers, be it in a community, workplace, and other human gatherings, are visited. Fictitious names and locations are used to protect the identity of perpetrators. Should you find some characters and behaviors resembling your own or someone you know, perhaps it may present an opportunity for some introspection, self-healing, and self-correction. The purpose of this chapter is not to indict or point accusing fingers at people, instead to highlight the practical and real-life scenarios and circumstances that the new entrant endures. By recognizing and understanding these situations, the newcomers are better prepared to face the hand dealt them.

Conversely, the stories might evoke some personal examination on entrenched citizens who may be deliberately or unconsciously engaging in these antics and behaviors. Unfortunately, these circumstances reflect the real operative environment of the new arrival. Below is a collection of stories told and recounted by some newcomers. These stories might assist new arrivals and entrenched citizens in understanding and appreciating each other and perhaps usher in a more friendly community.

Let us start with Bob, born and raised in a semi-rural community in the southern United States. Bob had never traveled outside of the U.S. The farthest he had traveled within the U.S. was during his high school excursion to the nation's capital, Washington D.C. Bob's experience and worldview are somewhat limited. Bob prides himself as patriotic and a dixie product, he would often proclaim. Bob does not wish harm on anyone and never misses church on Sundays.

Where Bob struggles tremendously is on his belief that the proverbial American pie is static and unyielding. With that mindset, Bob sees and interprets newcomers to this land as infringing on the fixed pie. It cannot be enough for everyone; Bob would vociferously argue and lament. This ingrained notion compels Bob to be antagonistic towards whomever he perceives as biting off the pie. Bob is operating on what has been dubbed the scarcity mentality. He cannot imagine the pie as expansive or able to grow. Bob's fear and motivation are not driven primarily by hatred, rather the subconscious desire to protect what he believes are the earlier arrivals' prerogatives. Poor Bob. If only he realizes that his forefathers were once like the current newcomer. Bob unwittingly engages in activities, pronouncements, and behaviors that tend to deny the newcomer a solid foothold in his or her new country. He does not hesitate to bellow out to anyone he deems a newcomer, "Go back where you came from." Bob does not comprehend or refuses to understand that most of the new entrants he insists must go back are bona fide citizens of our great country who enjoy the same rights and privileges as him, no more, no less. He often complains about what he terms job displacement of the "real citizens."

Bob's mindset and inclinations can be psychologically and emotionally threatening to the new entrant. It is real and front and center. So, how does the newcomer navigate through the likes of Bob? I submit by seeking to understand what is behind Bob's stance and frame of mind. Where and when possible, attempt to engage Bob and portray your ultimate goal of being a positive contributing member of society. Model behaviors that are disarming of Bob, who would otherwise put up a combative posture. Do not expect an instant conversion to your point of view or perspective. Bob had formulated his mentality over several years, if not several decades. Bob's surroundings either directly or inadvertently reinforced Bob's worldview. Should you determine that Bob is incorrigible or hopelessly ignorant in his beliefs, you purposefully avoid Bob's interactions. If he is a coworker or neighbor, or even a member of your church congregation (which can be more prevalent than you think), be just functional with Bob.

Let us segue into the workplace. Heather told of her experience within a healthcare setting. Heather is brilliant and obtained a medical degree in her country of birth before immigrating to the United States. Upon arrival in America, Heather took refresher courses and passed her board examination. She subsequently obtained her license to practice medicine. Heather's ultimate goal was to establish her clinic. In the meantime, she decided to work in a large Healthcare system in the western region of the U.S. Within a few years, Heather's brilliance, competency, and hard work earned her some promotions that she became the head of her discipline within the network. Whereas some "sharks" within their network observed Heather's ascendancy, it became intolerable once Heather got to the apex of her position. Then the proverbial sharks came out of the woodworks and attacked. These were in the forms of sabotage, undermining her efforts, and just plain intolerance. Nothing is peculiar with Heather's story. It represents a typical workplace politics and scenario. Yes, but in Heather's case, most of the negative utterances and pushbacks were anchored in the belief that she was too new in the system to have garnered so many accolades and promotions. Observe that the senior management of the network cherished and greatly approved of Heather's performance and skills.

Heather was told numerous times by her detractors, the so-called American-born peers, that her accent should be disqualifying. Mind you that Heather's accent though slightly different from a typical U.S.-born measures adequately in terms of communicability. She never had difficulty communicating her thoughts, and no patient or other business associates complained about an inability to understand and comprehend Heather. So, what is the point in Heather's story, one may ask? It is simply that as a new entrant, there are many obstacles, both blatant and subtle, that litter your path. You can choose to ignore them at your peril, or you can work to understand them and handle them properly and effectively. They exist! In Heather's case, she initially focused on performing her tasks and responsibilities excellently and doing the people conversion gradually. By people conversion, I mean letting her superior work prove to the detractors that her primary purpose is to serve and serve well. Over time, Heather brought a majority (there were only four visible detractors) over on her side. All she clamored for was to be judged by her work and not by her ethnicity. Today, when Heather tells this story, she laughs with joy and appreciation for how she dealt with the hand she was dealt.

Let us discuss Jesse. Jesse, as he is popularly called, emigrated to the United States about twenty years ago. He had trained as a marine biologist and had landed a job in Florida. He and the then fiancée got married and decided to buy a house of their own. After searching for many months, they settled on a mid-size house in the suburbs. They were very excited that their American dream was gradually being realized. They moved into their new home. They invited all of their neighbors living within earshot of their house to a small introduction and housewarming party with an exuberant spirit. They dropped the invitations in each neighbor's mailbox.

Then came the day of the party. All invited neighbors showed up except the neighbors to their house's left. At the party, Jesse innocently commented that the neighbors to their left did not come. Two neighbors at the party looked at each other somewhat apologetically. Though observant of the body language, Jesse did not think much of it. After the party, one neighbor stayed behind while everyone had left. He summoned Jesse and his wife and began …. "I don't know how else to break it to you other than just blurting it out so you are aware and can adjust to it accordingly. Your neighbors to the left are beside themselves that you guys moved in here. They are dead set on making your lives miserable so that you would move to another location. We disagree with them on that, and they have begun to isolate us." As the neighbor reported this to Jesse and his wife, you could feel the blood rush out of Maria's (Jesse's wife) body. She felt sick to her stomach, so much so that she found a chair to sit on. Jesse managed to utter a question by asking why the neighbor to the left would conclude such a stance without even meeting them up close. The visiting neighbor stated it was due to their nation of origin. He elaborated that the neighbors to the left cannot stand anyone not born and raised in the United States. Another gut punch! Before the visiting neighbor left, he reassured them that they should not let this upset them much because they are welcomed and supported by most of their neighbors.

Jesse and Maria just got hit in their guts and for a reason that was beyond their control. Their nation of origin. Could this be true? The answer is yes and yes. Here are Jesse and Maria, who are very proud of their achievement and looking forward to continuing to pursue their American dream. They have committed on the house and cannot just be driven away. How did it all end up, you may wonder? I will spare you the wait. The neighbors to the left moved out of the neighborhood seven months after Jesse and Maria moved in. Jesse relayed that they heard they moved farther into the county's unincorporated area.

So, what do you do when this reality hits as a new entrant attempting to better you and your family's life? It is real. It happens, though thank heavens it is a somewhat rare occurrence. In Jesse and Maria's case, they could not uproot and move, mainly due to the sunk investment in a house. You hunker down and have minimal interaction and contact with the hostile and unaccepting neighbor(s). When communication is inevitable, remain as pleasant as you can humanly be. If you own pets, never allow your pets to wander into their premises or yard. Avoid all possible situations that could lure you into a confrontation with the intolerant neighbor. Should one brew, always and immediately involve law enforcement officers for resolution. One could drum up excuses to pick a fight with you and your family. Be smart and deliberate.

Do you think soccer moms are spared? No, not at all. Let us visit this soccer mom's experience. Anna was a proud mother of a middle school girl. Her daughter was a budding soccer player with some promise. Soccer was a popular and trendy sport in Anna's country of birth, and Anna dabbled in the sport as a young adult unsuccessfully. Much to Anna's delight, her daughter took to the sport, who ensured shuttling her to practices and games. At these practices and games, Anna met a group of moms who supported their daughters' interest in soccer and therefore committed to bringing them and watching the team practice and play scheduled games. Anna is by nature an introvert who is slow at making friends. She slowly came out of her shell and opened up to some of the soccer moms. She felt comfortable enough to hold casual conversations with some of the ladies. Anna thought that she had been accepted into this informal group. No reason to think otherwise, she surmised.

Then there was a match on a Saturday midmorning, and Anna diligently took her daughter to the game. After watching the first half of the game, Anna dashed to the restroom to ease herself. Unbeknown to the two soccer moms in Anna's informal group that Anna was in one of the stalls, they struck up a conversation about Anna. Their comments were not only uncomplimentary; they bordered on outright hateful and harmful swipes and tropes. According to Anna, what made the remarks so devastating was that they were based on her national origin. Statements such as: "I can't stand her, and such that I never would sit close to her", "If by accident we find ourselves next to each other, I would find an excuse to move", "I don't even understand her when she talks. I just pretend and nod my head. Her accent is so thick and horrible." Anna kept her silence in the stall until they left. She went home and relayed the episode to her husband, who consoled her to take heart and avoid that group of soccer moms. Anna confessed that the experience negatively impacted her self-esteem for some time. Anna reasoned that her demeanor was non-threatening and considered herself relatively quiet and amiable. She played the soccer moms' restroom conversation over and over in her head, and it kept revolving around their intolerance of her, mainly due to her ethnicity and place of origin. It is real. It happens.

How about the children? Oh yes, they are a part of the journey, or should I say saga. For newcomers with children, they are always in tow as the family migrates to a new land. As naïve and innocent as they are and should be, they catch their share of intolerance and hostility. The negativity is often meted by peers and classmates, who are either well "educated" by their intolerant parents or emulate some of their peers' behaviors. To illustrate this experience, I tell the story of Megan's children.

Megan and John had two children- an eleven-year-old boy and a nine-year-old daughter- when they emigrated to the United States. They were academically average students and well-behaved. Upon their arrival in their new land, they enrolled in the public middle school near their neighborhood. As with most children, adapting to their new environment was easier for them than their parents. They made friends quickly and soon were somewhat indistinguishable from the other well-entrenched kids. Things seemed normal and steady.

Then one day, the oldest child, alighted from the school bus that just dropped both children off near their residence, crying uncontrollably. Megan, their mother, who came to receive them, rushed to him, loudly inquiring what the matter was. After gathering himself, the boy reported that three children from his class verbally bullied him as they were boarding the school bus. Megan became even more curious and asked the son what the bullies did or said. It was then that the boy stated that they loudly told him that he did not belong here and should go back to his country! Megan's son continued by asking the mother that he thought that she and his father had told them that they are now citizens of the United States and should not only be proud but cherish and revel in it. Megan's daughter who had observed everything- the bullying as well as the interaction between his brother and their mother- chimed in to emphasize the brother's question by posing, "Yeah, mom, why are they saying that we don't belong here and should go back to where we came from?" Megan, who was flustered by the kids' experience, calmed them down as they walked a few hundred yards into their home. As a nurturing and protective mother, she sat both children down to explain that in this world, including their original country, there are usually a tiny fraction of bad people, and bad people are not restricted to adults. Young people can equally be hurtful, and you can even find a few in your school or playground. She concluded by advising their children to ignore, and if necessary, report any bullying or abuse to the school authorities. To provide higher confidence to the children, she told them that the ultimate government authority, the Federal Government, had granted the entire family the status of citizenship, which bestows the same rights and privileges as those bullying them. They cannot rescind their right to be here nor affect it in any other way. This assurance seemed to do the trick in restoring some calmness to both children. It is real! It happens!! It is a part of the new arrival's ordeal. It is an inescapable experience and hand dealt the newcomer, who must muster all needed skills, attitudes, and fortitude to navigate these inevitable experiences.

This next story has an ironic twist. I will tell it and, hopefully, you will see what I mean. Some have posited that the most poignant and wrenching, and perhaps hurtful insensitive behavior meted on a newcomer is the one from another newcomer. Chandra and his family's experience demonstrate this sad and ugly notion. Chandra and his family had emigrated from Nepal six years ago. They settled in Georgia with Chandra working at a pillow manufacturing plant, while his wife worked at a neighborhood grocery store. Their two children were in seventh and ninth grades, respectively. Life was good for the family.

One Friday evening, the family craved Chinese food for dinner, and they decided to eat out. They asked their high school freshman daughter, who had become the family concierge, to seek and choose a Chinese restaurant for the family's dining experience that evening. With Generation Z-technology precision, she quickly pulled up and evaluated some restaurants on her Iphone. As expected, the family consented to her pick, and off they went. They arrived at the restaurant in about eighteen minutes and were seated appropriately without delay. It came time to order. Their waiter, whom they claimed seemed Chinese and spoke Mandarin to other coworkers, came to take their orders. Chandra understood and spoke some conversational Mandarin. Before he got married, he worked in Shanghai for four years and learned the Mandarin language. Due to his limited proficiency in the language, he rarely spoke it.

Chandra was a very finicky eater and remained very selective on what he ate. Ask his wife, and she will immediately descend into a long story of the extent and degree of Chandra's pickiness with food. She dealt with this every day, so she knew. Everyone placed their order with the waiter except Chandra. Then, it came Chandra's turn. Before this time, Chandra had reviewed the menu and forethought how he would combine some of the menu items. Chandra's first question to the waiter was whether cross-selection of items was allowed, and the waiter unwittingly gave an affirmative answer. Chandra then went at it. He cross-selected and mixed and matched items to the order taker's utmost chagrin. According to Chandra's wife, you could see by his body language that he was very irritated. When Chandra was finished with his marathon ordering exercise, the waiter stormed away in obvious disgust. It did not end there. What happened next became the crux of this story. Chandra and his family were seated almost at the restaurant's rear, not far from the kitchen. As the waiter got to the kitchen, he burst into Mandarin, cursing Chandra and his family, not realizing that Chandra heard and understood every word of his outburst. And this is what he said, "That man and the family are stupid and idiotic. They are not even from this country and should just be happy they are here and just eaten whatever they find on the menu, rather than trying to recreate their home country's dishes here. Or better yet, they should go back to whichever rat-hole country they came from." Wow!

Let us catch our breath on this first. Here was a newcomer unloading on another newcomer on a matter triggered by insensitivity to one's cultural background. One wonders if this sentiment unleashed by the waiter would have been expressed were Chandra and his family been red-blooded Caucasian

Americans. We would never know. However, the waiter's outburst and reference to Chandra and the family's homegoing signified an attitude based on their perceived ethnicity and national origin. This story's moral is that intolerant and culturally insensitive attitudes and behaviors are not the exclusive provinces of entrenched citizens. Newcomers are equally capable and do indeed dish out these behaviors and antics on fellow newcomers.

So, what do you do? First, you cannot stop living because someone does not accept you for reasons beyond your control. You cannot obliterate your past and background to please someone. How would you go about doing so anyway? It is impossible. As a newcomer to the land, you must learn to function against these odds. In the soccer mom's case, to remain focused on your daughter's soccer learning and enjoyment, and in Chandra and his family's episode, enjoying your family dinner and outing. You must dig deep and cling to a loftier purpose for whatever activity you are engaged in during the disruption, be it at work, at a gathering, at school, etc. You cannot stoop to your detractors' level either. One can pretend that these stories are unreal or blown out of proportion. Unfortunately, I hate to relay that they are real.

Recognizing these societal dark sides' existence does not make one an alarmist or a person full of negative vibes. It is part and parcel of the newcomer's "badge of honor" by these occurrences coming squarely at you. It is suggested that perhaps anticipating these intolerant behaviors mitigates the shock and deep scar such might inflict on you. Call it preventive steps in the newcomer's journey. Naively expecting full acceptance by everyone you run into at all times is merely setting yourself up for a huge disappointment. You are more likely than not to be dealt the more realistic hand of a few bad knocks.

Before I move on to the next chapter, allow me to pose some general questions for all of us: native-born, settled citizens, and newcomers. First, for the newcomer: have you found yourself in similar predicaments or ones that go at the heart of one viewing you from the lens of "not belonging" to the community? If you are now comfortably settled in your new land, have you meted similar behaviors to arriving new entrants? Do you condone or look away when observing others exhibiting such intolerable behaviors?

If a native-born or earlier settler: have you found yourself either consciously or inadvertently radiating or practicing intolerant antics and behaviors? Have you attempted to assist any newcomer one way or the other? Do you get irritated when you may need to go out of your way to explain or understand a new entrant? Do you get exasperated when the person on the other end of the phone possesses an accent, any accent, southern, Brooklyn, foreign? Do you make a deliberate effort to avoid contact or interaction with a newcomer because it would require a little more energy from you?

These questions are not intended to elicit any guilt feeling or self-flagellation. Instead, represent a prod for some introspection and pondering. It is through self-evaluation that we learn and grow.

"A forgotten past is a past that is yet to be."
-A.E. Samaan

CHAPTER 9

FLASHBACK. REFLECTION ON OLD COUNTRY

It has now been several years, even decades, since emigrating to your adopted country. However, nostalgia and what was would not be vanquished from your mind and memory. Often, when a bad experience shook your core, you immediately reflected on how things were in your old country. However, such memory mining usually resulted in your quick retreat from this sentimental lane. This pullback is because you cherish your new land and would make a similar migration choice given a second opportunity. The freedoms, liberties, economic vibrancy, and relatively more organized society that you are enjoying are too great to sacrifice for return to yesteryears. You probably could not return comfortably anyway.

The niceties that you embrace and treasure in your new land do not prevent you from mentally revisiting your old land's good virtues. These experiences include but are not limited to the extended family network system. Many countries' cultures possess a strong and expansive kindred web. Such a network enjoys a formidable support system that aids both emotional and financial sustenance. Matilda, who emigrated from Africa, extolled the extended family network that rallied for her financially when she secured admission into a university. Because her parents could hardly sustain themselves economically, relatives organized and contributed towards her four-year university education. It is usually an unwritten understanding that one pays it forward, and Matilda did just that by initially sponsoring another extended family member in his university educational endeavor. Seven years after emigrating to the United States, Matilda visited her old country during the Christmas season. She spent two and a half weeks visiting her birthplace, town, and her paternal and maternal homes. She also visited some childhood friends. Matilda confessed that the visit awakened her to the real sense of despair and suffering by folks at her old hangout. It was a humbling trip, she maintained. During this journey and experience, she pledged to sponsor two additional relatives in university education. This generosity entailed providing required funds bi-annually towards this commitment over four years.

How about the freedom and air of safety and security one felt in the old country? Jonny lamented he missed the evening and nighttime gallivanting around the neighborhood as young adults without fear or concern for violence or any harmful experience. He told how they would make merry and revel in youthful exuberance with reckless abandon. What was interesting was that older generational cohorts understood and approved of such display and appreciation of youth. Parents would often encourage their young adults to engage with their peers in such a rite of passage. Most newcomers from that part of the world recount these youthful forays into life as their most positive, memorable experiences growing up. What made such experiences even possible was the relative safety of their environment. Crime and violence, though not absent, were very minimal. Folks, young and old, saw each other as their brother's keepers. This air was very pervasive even without any written laws and regulations. It came naturally, or should I say, bequeathed from generation to generation that it became seamless and effortless.

Respect, especially for the elders, was instantly earned through one's age. It did not matter what academic, financial, or spiritual credentials one had garnered. What propels you to the respected ranks is your age. Elders always held sway, and society usually weighed their ideas, utterances, and perspectives higher than the younger ones. Maybe one day, we can discuss the advantages and pitfalls of such an elderly-dominated system. The premise for such an "elder-knows-it-all" culture is anchored on the belief that through life experiences, both good and bad, the elder is seasoned to proffer wise, rational, and reasonable thoughts and decisions. Of course, that could not be true all of the time. Cultures that tend to value every generation's ideas and propositions excel more than those stuck in the so-called elder-dominant culture. One does not have to look far to observe that inventions, innovations, creative approaches to issues are not the province of older members of the community alone. All generational cohorts contribute towards progress in society. Nonetheless, the respect-hierarchy system is often remembered as a non-disruptive and non-threatening structure and order. It was more comforting and enduring, hence the nostalgia.

If you were a family with an infant or very young children in the old country, chances are you did not take them to the daycare center or, for that matter, hire a nanny. The firepower and resources for such chores reside in your home through live-in relatives and kindred folks. Aging parents lived with their adult children, who nurtured and took care of them in their twilight. Such was the reciprocal reverse-caring culture. In South America, the Caribbean, Africa, Asia, and some parts of Europe, this practice represents the norm. So, there may exist three or four generational cohorts cohabiting happily within a household. Sometimes, the amicable and symbiotic coexistence amazes those from other cultures. Knowledge and small skills transfers occur and thrive in this setting. Grandma and Grandpa always enthusiastically shared their experiences, wisdom, and life lessons. This congenial and family bonding environment was usually enriching and cherished. The newcomer misses this in her or his new land.

How about the exhilarating feeling of debt-freedom? With no or minimal credit availability, one was not saddled with bills, financial shackles, and debt. This financial freedom is because whatever one purchased was settled with cash. This system, though delayed gratification, caused people to live within their means. Little wonder that some studies concluded societies with lesser economic complexities enjoyed greater happiness.

In later chapters, promptness and total devotion and adherence to time and the American attitude towards the clock will be discussed. For the moment, I will preemptively state that most newcomers immediately feel the oppression of the clock upon their arrival. Work is time-based, appointments are strictly time-hinged, gatherings such as weddings, community meetings, and social parties are time-sensitive. Whereas in some cultures, these activities are equally time-based, the degree of compliance with exact times is significantly different. In these societies, time is only a guide and not an absolute construct. For example, a wedding scheduled for 2 pm would commence at 4 or 5 pm. And everyone is usually aware of this quirkiness with time, hosts, and attendees alike. Hosts and event organizers would therefore exaggerate a start favoring an earlier time. Where am I going with this, you might wonder? Well, enough of the context. Here I come.

After sojourning in your new land, the United States, and perhaps visiting your old country, this hitherto unnoticed practice becomes stark. The newcomer becomes nostalgic when the need to report at a gathering or meeting at one's pace and convenience sufficed. The repressive nature of time and the clock in your new land is magnified. However, something interesting usually happens. The newcomer would no longer endure the lackadaisical approach towards time and the clock in the old country. They quickly become irritated by the termed rude and disrespectful disposition towards time. For some, though, the visits present needed opportunities to "exhale" and relax a bit, timewise. A friend of mine relished his yearly visits to his old country. He enjoyed the opportunity to attend events and activities when he was personally ready and still fully participated in such events. This relaxed disposition was because he knew that vibrancy and engagement were still guaranteed whenever he got to the occasion. He observed that he would mentally readapt to his American reality with time and the clock after his visits.

Most cultures do not require notification or elaborate signaling and approval for home visitations. People easily visit each other without fanfare. One could be preparing dinner or stepping out of a shower when a visitor knocks on the door. The visitor is welcomed with open arms and adequately entertained. The casualness of visits to people's homes is an endearing virtue of some cultures. The promptness of such visits lowers the visitor's expectations. The host would comfortably declare that he or she does not have any goodies to share. The visitor would not be offended at all. They would sit around and hold a good conversation and dismiss. When Roseline visited her old country, she quickly revived this ritual. She giddily recounted how she went from one house to another in her old neighborhood without any notification. Every home screamed in excitement when they saw her. No appointments were required. At some of those homes, meals were quickly prepared as a welcome gesture. According to Roseline, during the long flight back to the United States, she pondered and appreciated this aspect of her old country's culture.

This chapter would not be complete without discussing the nostalgia of local food and cuisine. Authors Daniel Fessler and Carlos Navarrete observed that food is not just for nourishment, as it embodies identity. Food and cuisine are acquired through experience, exposure, and familiarity. Some of the food ingredients could be found in the new land, usually in dried or packaged forms, but it is never the same as the food prepared with freshly obtained foodstuff back in the old country. It takes a while for the newcomer's palate to adapt to their new land's food and cuisine. Eventually, the new entrant acquires the taste of the new land's food offerings. Stories abound of those who traveled to visit the old country only to observe that the then local food and cuisine never tasted as delicious as they once did when they lived there. Our bodies, taste buds, even biological and digestive systems alter over time.

Theophilus, popularly known and addressed as Theo, went back to his old country after four years of living in the United States. Theo was always a braggart with a big ego. He likes to strut and display higher taste in material things. So, it was no surprise that his visit to his old country was with a bang. In his new land, America, Theo held moderately paid jobs. He saved enough money for his trip. Upon arrival, Theo was eager and ready to demonstrate to those left behind that he had begun to attain his American dream. He was doling out cash to any relative or friend who asked or deemed needed some help. He treated his childhood friends to daily restaurants and barhopping, footing all the bills. Theo's trip's mission and goal were to manifest wealth and upward economic mobility. In one of the barhopping junkets with his childhood friends, he bought drinks, any kind, for all bar patrons while he and his friends were at the bar. His childhood friends were in awe and astonishment at what Theo had accomplished in such a short period. So much so that his friends regretted not emigrating to a new land searching for the golden fleece. They did not realize that though Theo was relatively better off, his ostentatious display did not match reality. In his new land, Theo was just a regular guy who strenuously paid all of his bills monthly. Theo's actual situation was evidenced by his struggle to pay his accumulated bills upon returning. It took him additional six months to catch up on all of his financial obligations.

The ever-present tension and tug for the male and female newcomer is the eventual choice of a life partner in marriage. Amos's story, who had staked on taking a bride from his old country whenever he was ready, amply illustrates this quandary. Amos had emigrated and lived in the United States for eight years. In one of his visits to his old country, he had planned to "look around" for a possible future bride. He went, he saw, and he came up empty. Amos had become westernized to no small extent in emigrating to the United States and acculturating in his new land. What could be wrong with that in his bridal pursuit, you may wonder? Lots.

Amos's perspectives, worldview, and perhaps, inclinations and behaviors have changed. So, he was told by relatives, he relayed. So, place this newly minted Amos against the backdrop of potential brides from his old country, and you have a problem. It goes both ways in that the potential brides would view Amos as compromised or even too jaundiced for their liking. While there, Amos took three different females on dates, of course, at different times. He was disappointed and dismayed at each outing. His reasons ranged from "they think as though they are from a different planet" to "I could not connect or have simpatico with any of them" and "they seemed more materialistic than I can stomach." You see, the problem was mutual. Amos had changed. Amos had become a hybrid. Amos had morphed into "Ameri-Amos," A blend of cultures from the old country and his new land. He expected this metamorphosis, so the tension he felt was realistic and practical.

That was not all. Amos had lived in the United States, where he enjoyed an organized society, nice roads, and an efficiently functioning system. He had become accustomed to these basic American features and luxuries. During his first visit to his old country, the broken system's starkness was very pronounced. He complained the roads were full of potholes, and the drivers, all drivers, drove like maniacs. Going and obtaining services from the bank was almost an all-day, at least half-day event. According to Amos, if all the systems in America were moving at jet speed, things operated in his old country at a snail's pace. Okay, a bit over the top, but you get the idea.

Interestingly, while Amos lived in his old country, he did not observe this significant disparity in systems. Well, how could he? After all, it was the only system he was exposed to all his life. Even the political and judicial systems received swipes from Amos. Having experienced superior political and judicial machinery in the United States for some years, Amos saw the flaws in his old country's processes. It became evident that the old country's systems needed retooling and enhancements. Amos was handed a reference platform for comparison by migrating to a new land, and the differences became evident and glaring. One of the offshoots of migration, especially to a better society, is the appreciation and cherish of your new land and community. Amos confessed his engagement and appreciation for his new land increased exponentially after his first visit to his old country.

Jaya emigrated from Indonesia and, after seven years, traveled back to visit relatives and friends. What stood out for Jaya upon returning to his old country was the different uses of motorcycles in both worlds. In the United States, motorcycles are used primarily for leisure, while in his native Indonesia, they are utilized mainly for utilitarian purposes. They are used to carry goods to unmotorable terrains and taxis, shuttling residents around cities and rural and remote villages. Motorcycles were more commonly used than automobiles due to their ease of maneuverability around tight areas and during road traffic jams. One could easily flag down motorcycle taxis within minutes of desire. He then clearly saw the vital economic tool that motorcycles represented in his native country. This concept was not as vivid to Jaya when he lived in his native land. It became a reverse-cultural shock for him when he visited his old country. Upon return to the United States, his new land, he never viewed motorcycles the same way. When he saw a leisure rider cruising down the highway, he mumbled under his breath that the motorcycle could be a productive economic tool that could feed a family in his old country. The irony of it all, Jaya owned a motorcycle for leisure in his new land, America. He had contemplated sending his leisure motorcycle to his cousin in Indonesia for use as an economical vehicle for livelihood. The shipping fees were exorbitant, so he scrubbed the idea.

For many newcomers, whose old country's systems and way of life may be archaic or still stuck in a medieval mindset, exposure to the new land's perspectives and approaches could be transformational. These newcomers undergo some cognitive metamorphosis of their worldview. For example, one sees and appreciates gender equality held and practiced better than in one's old country. More work and recognition are still needed on gender equality in the United States. However, comparatively speaking, America is ahead of some nations and societies on this subject. Emigrating from lands sorely behind in this sphere magnified the United States' progress so far for the newcomer.

Take Elena's case, who had emigrated from Eastern Europe roughly fifteen years ago. In Elena's old country, women did not enjoy all their male counterparts' liberties and respect. Yes, their voices were "heard" but often not taken seriously. They were regarded as a tolerable portion of society. Men dominated and ruled. Elena's sojourn in the United States changed all that for her. No matter how imperfect the American society, Elena's education and exposure to gender equality empowered her on this subject tremendously. When Elena first visited her old country after living in the United States for eight years, she became appalled and critical of how women were viewed and treated. She confessed that while she lived in her old country, she never fully realized how backward and ancient their societal viewpoint on women and their rights was. She credited her education and exposure to a different culture for broadening her horizon on this matter. Elena did something about her new-found empowerment. She saved enough money, and after two years from her initial visit, planned and established a women's center in her town. The women's center's purpose and mission were to educate and empower young girls and women in her old town. Annually, she would send funds for the operation of the center. She conducted fundraisers in the United States to generate the needed financial resources. In addition, she obtained donated old books, magazines, and journals from libraries in and around her city in the United States and sent them to the center. Elena maintains that the women centers' projects and associated activities provide her with one of her life's purposes and dreams. The women's center has been operational and waxing strong for more than six years. She makes an annual pilgrimage to her old town to monitor and guide the center.

Pedro's experience when he visited his old country in South America was shocking and troubling. He had lived in the United States for twelve years before his trip. He maintained that he braced for a lesser egalitarian society and environment. He had not adequately prepared for what he saw and experienced. The society and nation he left had changed significantly on one aspect. Its citizen's drive and effort to earn their living lawfully had taken a beating for the worse. Greed, avarice, and the urgency for quick wealth at all costs had dominated citizen's collective psyche. This perversion had become accepted and condoned at all levels of society. Illegal drug manufacture and trade had become the new career and aspiration of the younger generation. So much so that education had taken a back seat. Beautiful houses and fancy foreign vehicles were everywhere.

In disgust, Pedro attempted to reason with family members and former schoolmates regarding the moral decadence precipitated by the attraction to illegal drugs as a means of livelihood. He was unsuccessful. Most of the people he had talked to were directly or remotely connected to the sordid drug business. Some were farmers who grew the crops. Others were manufacturers who refined the products, while the rest were distributors and traders. Pedro was confronted with the reality of the conspicuous lack of employment opportunities within the local economy. For a moment, his thought drifted towards rationalizing and perhaps justifying the inevitable embrace of the illegal drug enterprise due to economic hardship brought upon the populace with no legitimate escape route. He quickly exited this line of thought. After all, poverty could not represent an excuse for breaking the law, he concluded.

Pedro's on-ground experience shattered his prior belief and faith about the economic rejuvenation of his old country. This hopelessness stemmed from the understanding that a society relying on an activity, such as illegal drug manufacture and trade, can never be sustainable. The economic engine of a community must be built and maintained by legal businesses and trade.

"*A people without the knowledge of their past, origin, and culture is like a tree without roots.*"
-Marcus Garvey

CHAPTER 10

MAIN PILLARS OF AMERICAN CULTURE

This chapter is devoted to examining the main pillars of American culture. These central tenets are discussed to assist and aid the newcomers in their yearning for assimilation into the new land. Also, it may represent a refresher for the native-born. The author will cast no judgments or aspersions on these cultural underpinnings' merits. We have learned through human history that there are no good and bad cultures. Instead, cultures tend to be relative to their surroundings and environments.

Philosopher Immanuel Kant in 1788 reasoned that the material and social world is mediated through our minds. Peoples' experiences of the universe are based on the knowledge and ideas they possess. An African proverb counsels that "The stranger sees only what he knows." It is impossible to identify universal experiences that resonate the same for everyone. Consequently, context remains the catalyst around our values, beliefs, and practices. Therefore, tolerance and respect for cultural differences must exist. The culture that is treasured in one society could be rejected in another.

Social scientists coined this notion as relativism. Cultural relativism refers to not judging a culture to our right or wrong standards, strange or normal, acceptable, or intolerable. For example, one cannot judge and dismiss as abnormal or weird, a culture that eats fertilized developing bird egg embryo, usually duck, incubated for about fourteen to twenty-one days, then boiled or steamed, with contents eaten directly from the shell. The consumption and enjoyment of this food item are connected to society's cultural background. In another culture, fried crickets are a staple. Does the high protein content of this insect influence its food quality attractiveness and adoption? We may never know, but the adopters most likely pinned their rationale to their environment and culture. Usually, specific and available ingredients are utilized to evoke desired flavors that capture the essence of the community. For example, the degree and intensity of spices such as pepper vary among communities and cultures. Often, correlations could be drawn between communities' food culture and their geography. Observe the riverine community's food reliance on sea creatures and lowland vegetation while dryland diets depend on available game and flora and fauna.

In some cultures, certain food items are reserved for adults. This practice has been traced to the scarcity of a particular item. Children are discouraged from developing the taste and appetite for such specialty entrée. A case in point is the chicken egg. Even with its rich nutritional value, it is stigmatized as an errant craving and a character flaw with the young. Children are therefore conditioned to suppress their interest in eggs until adulthood. That community's food culture was borne out of its surroundings and norms – an illustration of cultural relativism emphasizing food.

In another example, polygamy is practiced in some African cultures. It is an accepted custom that allows a man to marry as many wives as he desires and can support. Most westerners focus on sexuality than the social and economic reasons for such practice. Polygamy in these cultures is rooted in economic viability and productivity. These agrarian economies required many field hands in cultivating the land and tending to the livestock. Besides, a larger household permitted organized and efficient hunting and the division of labor in food processing and production. Furthermore, having more wives and children symbolized status and wealth and provided broad political alliances and networks. Larger families engendered protection from aggressors in the community.

As these cultures moved from dominant agricultural to light industrial economies, the need for field hands decreased. Consequently, polygamy declined and continues on this downward trajectory. Without a better understanding of the societal setting, the wrong judgment could be rendered on the practice of polygamy in these cultures. Therefore, cultural relativism must be at the core of any cultural evaluation and appreciation.

Relativism extends beyond culture and permeates to language. Anthropologists Edward Sapir and Benjamin Lee Whorf theorized on linguistic relativism. They inextricably linked language to culture. Certain thoughts in a language could not correctly translate to another language without losing its context and integrity, they posited. Culture and worldview create our vocabulary. Conversely, our thought process is shaped by our native language. To fully appreciate a language, its cultural context must be understood. From cultural settings evolve language nuances, slang, meanings, and interpretations.

I will digress and share a television watching experience that supports linguistic relativism, albeit among animals.

I was watching a National Geographic series on predators and survivors in the animal kingdom. What struck me was that as a group of animals was either feeding or gallivanting, they communicated a specific noise (language) that warned of nearby predator danger. Every member of the threatened family heard and accurately interpreted the information while the predators seemed oblivious of the language. Likewise, humans can hear the same language but interpret them differently based on their cultural background and worldview. Mariam's story in chapter 15 illustrates this phenomenon.

One of America's intriguing aspects is how the immigrants arriving from different parts of the world have created a uniquely American identity. This identity is rooted in individual freedom, self-reliance, equality of opportunity, competition, the American Dream, and hard work. The Frenchman, Tocqueville, visited America in the 1830s and observed firsthand the avid self-reliance and individualism of a budding nation. The intertwined nature of self-reliance and freedom was stark. America believes one must be self-reliant to preserve one's freedom. This cherished self-reliance affords one the freedom to live out their lives any way one chooses. The American character is anchored in the belief that people should control nature, surroundings, and destiny rather than be left to fate. This spirit resulted in a goal-oriented, energetic, and progressive society. America is often termed an "open society." This terminology stems from the transparent nature of individuals, institutions, and organizations. The press is freest than in any country on earth, as it is referred to as the fourth arm of democracy behind the executive, the legislative, and the government's judicial arms.

The orderly transition and transfer of government after elections in America continue to be held up as one of the demonstrated examples of democracy at work. Regardless of how close, controversial, and bitterly fought an election is, transfer of power is guaranteed for the sake of preserving democracy. A case in point was the 2000 Presidential election between Albert Gore and George W. Bush that ended in a Supreme Court ruling. Once the court rendered the verdict, all parties accepted the ruling and peacefully transferred power. Some of the newer entrants to America marveled at the universal and unconditional acceptance of that judgment. This amazement was because chaos and crisis would have erupted in most other countries. Though not perfect, as the bastion of democracy, America remains the envy of the world.

One can submit that America's self-reliance bent was forced upon it by circumstances. Let us face it, America was a frontier nation. Frontier nations initially rely on themselves. They scrape, dig down, and beckon on sheer grit for survival. Australia represents another frontier nation. To compound things, the distrust of government and institutions compelled the American pioneers to abandon their original country, Britain, to set sail to a distant land to escape the trappings of tyranny, oppression, and subjugation. The British experience was rife with nobility, class system, and conformity. The early American settlers wanted to get away from such a mindset and way of life. What makes the American dogma of self-reliance unique and exceptional, where "self-made" is a badge of honor, is that most countries subscribe and practice the opposite. In those cultures, self-made often is viewed negatively as selfish, self-absorption, and not your brother's keeper. Self-made proponents are regarded suspiciously as opportunists at the expense of the overall societal good and welfare. The American experience promotes competition because she believes that competition brings out the best in people and free enterprise results in successful outcomes.

Self-reliance embraces and advances self-worth, self-expression, self-knowledge, tenacity and doggedness, self-definition, and self-appreciation. It encompasses thinking independently, emphasizing individuality, dogmatically pursuing one's goals and purposes. It is referred to as "paddling your own canoe" in American expression and vernacular. The American experience espouses the non-permanence of one's condition. One can start poor but can excel to affluence and vice-versa, depending on personal efforts, hard work, and decisions. The adage of "rags to riches" is as American as apple pie. Little wonder the crafters of the American Independence declaration proclaimed that "...... all men are created equal and are endowed by their creator with certain inalienable rights, that among these are life, liberty, and the pursuit of happiness."

This section of this chapter may be deemed boring or even unnecessary. However, for the sake of providing some context and as a refresher, I have included it. Freedom in the American context is the product of self-reliance. The early settlers banished the restrictions and controls by established British institutions such as the monarchy, governments, churches, aristocrats, and noblemen. They rebelled against this system as they yearned for one that placed the power to govern on the people. The ardent need for freedom from these institutions' shackles prompted the American constitution crafters to separate the church and the state. This demarcation ensured that churches and religion were not sponsored and dictated by the government. These early settlers deliberately instituted a limiting power to governments and aristocrats, thereby creating and sustaining a system where the individuals' reign. This concept of individualism is interchangeable with freedom in the American lexicon.

Americans are generally independent, individualistic, and cherish their freedom. This independence streak breeds competition where the best rises to the top. The belief in freedom of choice runs deep in America's ethos. A clear illustration of this dogma was during the COVID-19 pandemic of 2020. A robust debate ensued on the wearing of face masks. Where science and data had determined and proclaimed that wearing masks and social distancing prevented or at least slowed the spread of the virus, some Americans resisted wearing masks to exercise their freedom of choice. Those that defied a public appeal to wear masks voluntarily argued they were exercising their freedom of choice rights. This argument was against the backdrop of public safety concerns. Some states lifted previously imposed mask mandates but strongly encouraged mask-wearing. These seemingly contradictory directives deference America's rooted belief in freedom of choice, even as the pandemic continues.

Contained in the Bill of Rights in the Constitution of The United States of America are five freedoms. Freedom of Religion, Freedom of Speech, Freedom of the Press, Freedom of Assembly, and Freedom of Petition.

Freedom of Religion: This guarantees all Americans the right to practice any religion they choose or to practice no religion at all. Congress is forbidden to establish, favor or tax citizens to support any religion. One often hears this referred to as "the separation of church and state."

Freedom of Speech: This is the right to express your ideas and opinions and the right to listen to others' thoughts and opinions. One cannot use this freedom to injure others. Should that occur, the harmed may sue in court and obtain a judgment for damages. Furthermore, one can express their opinion about their government under this freedom.

Freedom of the Press: This pertains to freedom to express your ideas and opinions in writing. This freedom is related to freedom of speech. Besides, it includes the freedom to read what others write. The writing may be in any form such as books, newspapers, magazines, and of course, these days, the internet.

Freedom of Assembly: Freedom of assembly involves the freedom to hold meetings, which are expected to be conducted peacefully.

Freedom of Petition: This refers to the right to ask your government to do something or refrain from doing something. This freedom gives you the right to write your representative and express your opinion and desire regarding certain laws, rules, and regulations. This provision represents an avenue for government officials to know their constituent's stance on certain items.

Many other freedoms and rights mentioned in the Constitution include Freedom to live and travel anywhere in the nation; Freedom to work at any job for which you qualify; Freedom to marry and raise a family; Freedom to receive a free education in good public schools, and Freedom to join a political party, a union, and other legal groups.

All of the freedoms in America's Constitution manifest and magnify the system of government envisioned by the founding fathers. This system, commonly referred to as a democracy, recognizes the government as the authority established by people to help them run their affairs. This government provides services that citizens acting alone could not perform themselves. This government is based upon the approval or consent of the people who are governed. Some powers reside with the federal government, and others are left to the states, while the federal government and the states share still others. This system of government is woven together in the three branches of government, namely, the legislative, the executive, and the judiciary.

The American culture of self-reliance, individualism, and hard work dovetails into the term and concept of the American Dream. The American Dream is rooted in the pursuit of personal fulfillment and the desire for self-actualization. The American Dream is the national ethos of equal opportunity for all. It determines one's success and outcomes based on one's sacrifice, risk-taking, decisions, and hard work rather than by chance. People have equal opportunities, as people are viewed and interpreted as individuals rather than their lineage or pedigree. Whereas opportunities may not be equal for all, compared with other societies, nations, and economic systems, the American experiment represents a superior apparatus for equalizing opportunities. The economic system rewards indiscriminately and contingent on what one brings to the table. Material goods are perceived and held as rewards for hard work and evidence of favor from God.

Throw in consumerism and you complete the American culture's fundamental wholesomeness. The consumerist culture refers to the acquisition of consumer products and services, such as cars, houses, clothes, devices, restaurant dinners, cable services, hair stylists, etc., instead of savings and investments. Consumers purchase these products and services to keep up with trends and the pent-up desire to upgrade products and services continuously. In the classic American economic model, consumption is emphasized and espoused. Crafters of this economic model positioned spending as the growth engine for a robust financial society. Its fuel is the credit system that allows for immediate enjoyment and gratification of goods and services with delayed payment. Institutions and organizations such as banks, credit houses, transactional enablers, etc., have all been built to support and sustain this economic model and system. Consumerism and capitalism have been deemed parallel and indistinguishable.

Consumerism is alive and well in America. The proverbial "keeping up with the Joneses" is as American as baseball. One can safely submit that this attitude, mentality, and way of life fuels the American economic system. You may change your well-functioning car because your neighbors have purchased two different vehicles while you tend to be clinging to your clunker. Your neighbors' kids all sporting the latest Michael Jordan tennis shoes, and by golly, your kids deserve those also. Your coworker, whom you despise and earns considerably more, is spotting the latest iPhone watch, and you cannot wait to get off work to stop at an Apple store to purchase one. These examples reflect the nexus of self-reliance, competition, and consumerism in the American culture.

American exceptionalism, or the perception of such altitude, powers both individual and national disposition and pride. American exceptionalism represents the belief that its history is inherently different from other nations. It is anchored in the notion that the United States is uniquely virtuous. This superiority complex has been echoed over the last two centuries through phrases such as "Shining City on a Hill," "Last Best Hope on Earth," "Leader of the Free World," and "Indispensable Nation." While this remains largely true, such chest-thumping plays to our nation's ego. It boosts our self-esteem as a country and a people. Nothing wrong with that. Every nation clamors and establishes its collective morale-enhancing spirit. Some have argued that the concept of American exceptionalism is only but a myth. But it does not matter if a myth or not. Should this rallying cry and belief serve their purpose of elevating the collective national superego, then well done.

Some of the components often associated with American exceptionalism include the institution of democracy and its principles. America remains the citadel of democracy and democratic institutions the world over. Though imperfect, it represents the closest testament to true democracy. The populace has a stronger and broader say in their lives. The people can influence their destinies, comparatively speaking. The people can choose their leaders. The people can determine how they are governed by the laws and policies they espouse through their representatives. Another area of touted American exceptionalism lies in inventions, innovations, and advances in human endeavors. America's ingenuity remains the envy of nations. The freedom enjoyed, the rugged individualism practiced, and the openness and transparency espoused all confluence powerfully to unleash creativity and inventive spirit. Scientists can work in their laboratories without interference. The artists can draw in their studios in quietness. The writers can put their pens to paper or type away at their computers with abandon, and the athletes can perfect their skills without fear. Such is America. A land where you are free to be whoever you aspire to be.

An aspect of American's way of life that is rapidly impacting citizens' health and well-being is our obsession with largeness, including food portions. I discuss this in greater detail in Chapter 11. Consequently, America has an obesity problem. This challenge represents the axis of our societal appetite for largeness and the abundance and variety of food items. Consider the following. According to the Center for

Disease Control (CDC), the prevalence of obesity in the U.S. was 42.4 percent in 2017-2018.

Furthermore, the estimated annual medical cost of obesity in the United States was $147 billion in 2008. Obesity-related conditions include heart disease, stroke, type 2 diabetes, and certain types of cancer. Obesity has slowly been accepted in the American psyche. Consider that soft drinks are cheaper than healthier bottled water. Free soft drink refills are the norm at restaurants. Fast food tends to be less expensive than healthy food. These adaptations signaling, albeit unintentionally, that bingeing, especially on unhealthy food and drinks, is acceptable. Any newcomer immediately observes this very American norm.

Studies have reported some correlation between the more economically developed societies and the level of obesity. These advanced economies are more likely to eat processed foods compared to less developed communities. Some have laid partial blame on newcomers, who cling to their ethnic foods mostly packed with calories. These food choices are familiar to them and comforting. Another school of thought points to the mixed messages American consumers receive. On the one hand, fast and quick meals are celebrated, big portions are tolerated, and processed and packaged foods are accepted. The weight loss and diet industries surreptitiously encourage us to splurge and come to them for remedy. Of course, with our wallets. We listen but cannot stick with the regimen to shed gained pounds. Then we repeat the same process annually.

In some cultures, skinny is not chic. Plump and fatness are craved. For example, in some Middle Eastern and African cultures, when females get married, they are sequestered and overfed in the so-called "fattening rooms" to look their best for their new life. In these cultures, overweight is a sign of good health. Skinniness is interpreted as negligence by and poverty of your husband. However, with the proliferation of technology and digital devices, cultures are emulated worldwide. The younger generations now shun fatness for slimness as the definition of beauty and elegance.

A revered artifact, the Venus of Willendorf, (dated between 25,000 and 27,000 years) that was discovered in Austria in 1908, depicted an anatomically fat woman. Europe and Asia boast of Venus figures from the Stone Age (50,000 to 10,000 years ago). All of these figures are overweight. I resist speculation here, but it is intriguing to observe these fat figures. Could it be that fatness was celebrated and heralded? We do not know. Scientists caution that obesity and plumpness are different. Moreover, who sets the yardstick for body types and weight measurements? Should a universal marker exist or be left to each society? I am unqualified to wade into this debate, so I will leave it alone.

Often, when you are new to an environment, you are more apt to observe your previous society's differences. On this note, most newcomers appreciate and admire the American business' customer-focused orientation. Efforts and activities are always geared towards attracting and maintaining customers. The customer's convenience and comfort are paramount, and a good value proposition and buying options are presented. This need to positively lure the potential customer is at the heart of America's economic model. Turn on regular or cable TV and watch the advertising deluge experienced. These inducements are aimed at enticing potential customers. A business perishes with inadequate attention to the current and would-be buyers of goods and services.

An acquaintance who had emigrated from the Caribbean sensitized me to this occurrence at some non-native-born-owned independent stores. These stores include mom-and-pop, gas stations, and other utility marketplaces. He had submitted that these stores are generally less focused on treating the customer as "king" or "queen." The example he cited was requesting the use of restrooms at such establishments. You are seen and treated as a nuisance and begrudgingly shown the restroom. Often, these stores frown at just utilizing the restroom without any purchases. This perspective is devoid of the concept that the restroom user represents a potential customer of their wares and services. My acquaintance might have ruined it for me because I observe this snubbing more frequently after drawing attention to the behavior. I engaged in an unscientific survey by asking two storekeepers at different cities about their feelings upon request for using their restrooms to satisfy my curiosity. The responses were both emotional and intense. One reasoned that the stores should never represent off-ramp easing facilities for universal use. After all, he continued, "We pay people to clean them and get nothing in return." He finally stated that were these provisions not mandatory, he would not install them for public use. The second respondent, though less strident, was equally resistant to the required furnishing of the amenities. They both were oblivious of the possible patronizing of their businesses by casual restroom users. Their inadequate customer or potential customer focus triggered using their restrooms an irritant.

The prevalent perspective and practice in some cultures are where business establishments view themselves as providing needed services. Therefore, the consumer ought to appreciate and reward them for their efforts in furnishing these goods and services. Little wonder these stores project less concern for the customer. In these societies, the scarcity of products and services makes their availability valued and cherished. Businesses, therefore, lack the motivation and incentive to cater to the customer. However, in America, goods and services are abundant with various options and preferences. This reality of many consumer choices breeds marketplace competition, including courting potential customers.

The conglomeration of the different and several parts of the American values milieu portrays American identity. They remain the sum of the parts that make America, America. The inseparableness of self-reliance and freedom, the inextricableness of hard work and doggedness, the connectedness of equal opportunity, and the pursuit of one's goals lead to the American dream.

*"A love of tradition has never weakened
a nation. Indeed, it has strengthened
nations in their hour of peril."*
-Winston Churchill

CHAPTER 11

OTHER CHARACTERISTICS OF AMERICAN CULTURE

Many American values and culture components are pronounced and vivid, while others are subtle and subdued. Nonetheless, each is important in its own right. We will examine many of these attributes, but this discourse does not represent an exhaustive list of characteristics.

We established in chapter 10 that America remains the melting pot as it harbors a diverse group of ethnicities, practicing many religions, and speaking multiple second languages. This feature though advantageous can be wrought with potential setbacks, thereby stunting progress in society. For example, because the various ethnic enclaves are distinct and largely separated from each other, it helps preserve the ethnic culture but prevents or delays the complete assimilation into the mainstream American culture. Nonetheless, the main characteristics of the American culture and way of life exist. Let us explore them.

Americans need and must have their elbow room. This inclination describes the personal space around the individual. In some cultures, this personal space is limited and smaller. The restricted space usually demonstrates closeness and a degree of comfort between individuals. Notable cultures in Eastern

Europe and the Middle East practice smaller personal space as people interact. When Americans travel to such regions, they always tend to be pulling back from whomever they are interacting or communicating with, creating their needed space.

Interestingly, people from cultures with constricted personal space often interpret the pull back as a disrespectful or disapproval body signal. Citizens do not push or stand too close to anyone in line. They always wait their turn. Jumping in line remains non-existent in the American culture and way of life. Again, this mindset corresponds with the personal responsibility and accountability bent inherent in the American culture.

Americans respect authority and follow the rule of law. This tendency remains one of the sustaining tenets of democracy. They also generally do not argue or banter with police and other law enforcement personnel. They simply abide by the laws and regulations. Examples abound of the manifestation of this phenomenon. Take, for instance, the organized and systematic obedience of traffic lights or the adherence to no jaywalking. Chapter 2 discussed Cecelia's marvel regarding the strict compliance to traffic signals upon her arrival in the United States.

Another example is the littering, graffiti, and tagging discipline exercised by the citizens. How about the pedestrians having the right of way and not being mowed down by motorists, or the control exhibited in not loitering in areas? To buy or drink alcohol, you must be over the age of 21 and required to show a photo identification card. In most states, it is illegal to purchase cigarettes if under 18, and smoking is restricted or not permitted in certain establishments.

Laws and regulations are enacted to protect the vulnerable and the weak. For example, the law prohibits discrimination against people based on their national origin, religion, sexual orientation, etc. In certain cultures, including those from whence the newcomers emigrated, behaviors such as same-sex intimate relationships are prohibited by law and, in most cases, punishable by incarceration. A sexual orientation that differs from the societal norm is considered deviant, and therefore not acceptable. One can successfully submit that America's culture of personal responsibility and accountability again fuels the entrenchment of these cultural foundations.

The American culture shuns formality. Formality, therefore, is interpreted as arrogance and demonstration of superiority. This penchant for casualness produced an informal, egalitarian society. An acquaintance who migrated from New Zealand in 1994 lamented not addressed as "Mister" or "Boss" at his job. He complained that the employees were not addressing him with a title. They disrespectfully called him by his first name, he protested. He claimed he later resigned to this embedded American culture of informality in addressing whomever by their first name. For most newcomers to America, this practice represents one of the initial cultural shocks. Imagine being addressed as "Doctor," or "Mister" or "Madam" in your old country only to be seemingly reduced to your first name or nickname in your new land. It takes getting used to, and often not easy. Adrian, who migrated from Europe, told how he was bothered and angered by strangers who would immediately engage in small talk at the drop of a hat. "At line queues or small gatherings, people, strangers would not leave you alone," he grumbled. You see, Adrian was hopelessly introverted. So much so that he abhorred and avoided unnecessary interactions with others.

He could happily live and survive in his cocoon. Therefore, imagine being accosted by and spoken to by strangers at every turn. Welcome to America, where small talk is regarded as extending a friendly gesture and representing societal norms. It embodies her citizens' affability disposition and is applauded in part for helping newcomers' overall assimilation. Today, Adrian can now talk your ears off and does seamlessly and shamelessly speak to strangers comfortably. Friends are currently teasing him as "Americanized."

The chances are that if you stand in line at a store or other queuing circumstances, you are more likely to be spoken to by a stranger on the line. Adrian's example above exemplifies such occurrences. Such casual conversations will often lead to learning about the stranger's condensed life history. This openness is because in America, people talk easily to strangers. This ease and relative comfort in information sharing and disclosure epitomize America's open society. Personal, institutional, and government transparency is not only enjoyed but demanded. Here comes the oft-seeming contradiction. While openness and transparency are cherished, privacy is equally revered. However, upon a closer examination, they are not contradictory as it initially seems. There are certain areas and topics Americans hold private. These include age, salary, their political, sexual, and religious views. Everything else is fair game.

Before we move away from the American informality and casualness culture, allow me to share a personal experience about this aspect of the American culture. Upon arrival and enrollment in my college so many decades ago, I had a close encounter with this culture. On the first day of one of my elective courses, Music, I was seated and nervously ready for class. Then strolled in this gentleman who could have easily had come straight from the sixties Woodstock Music Festival or a hippie commune: long hair and beard to match, shorts, and a flannel shirt. As if he was determined to take it to another level, he had a pair of bathroom slippers. He walked towards the professor's desk and chair and announced loudly, "Good morning, I am Doctor Peck (not real name), your instructor, and we will be working together in this course for the next several months." I was extremely shocked. I am sure that if anyone recorded that moment, they would catch me with my mouth agape. I could not fathom a professor coming to class and holding a session in such casual attire. Yes, it was spring, and the weather was starting to get warm, but, by golly, not that casual, I reasoned.

To provide some context for my shock and disbelief, college students dress their Sunday-best to attend classes in my old country, and the professors would not be outdressed by the students either. So, to have seen a professor dress so exceedingly casual, as I termed it then, blew my mind. My incredulity was that the professor was unfazed and focused on his lecturing as though there was no problem at all. There was no issue here but for my jaw-dropping sight. I was the one experiencing bewilderment and an "antacid moment." I looked around the classroom, and no one seemed to be bothered, at least outwardly, by the professor's appearance. Most, if not all, of the other students were native-born Americans. The professor completed his lecture an hour later and left. I am sure that I did not learn a thing in that class that day. I could not concentrate as I was consumed with what I was witnessing. It was indeed an eye-opening experience that stayed with me. He was one of the most popular professors on campus because he mastered his subject area and taught in an enlightening, relaxed, and enjoyable manner and atmosphere. "Welcome to the U.S.A.," I said to myself later.

Let us discuss the so-called "going Dutch." This term refers to the practice of each diner paying their share of a bill or splitting the bill in half. This practice applies to all regardless of the gender of the diners. Some men from other cultures often feel threatened by such moves and interpret this behavior as too independent for their liking. Most American women are not comfortable letting men pay all of the bills. The discomfort of mostly men from different cultures is rooted in men's regard as the providers. While in college, most men from far away cultures debated and concluded that most American women who insisted on paying their restaurant bills when they both went out for dining depicted extreme independence, which was a loud siren for a potential culture clash. They would count such assured behavior by their date as a good reason not to pursue such a relationship deeper. To them, it represented a microcosm of some deeper irreconcilable cultural fissures.

On the same corollary, it is normal for American women to maintain male friends who are just friends and vice versa. Any such friendship arrangement in other cultures is virtually nonexistent. Indeed, it is standard to find American men and women sharing an apartment as roommates and no sexual relationship attached to such cohabitation. Also, in America, women can propose to men in heterosexual marriages. Though the percentage of such proposals is low, reported at 5 percent in 2010, but gaining ground and recorded at 16 percent in 2019. Some cultures vehemently shun this practice.

In America, time is money. Lateness to work, school, appointments, parties, weddings, and even funerals is frowned upon. Little wonder hourly and most types of work are time-based. In some cultures, time is a relative concept, and punctuality takes a backseat. Lateness is expected, thereby compelling people to state fake times in advance of events, realizing that attendees would not show up at the designated times. In some cultures, work, for example, is measured by the day and not by the hour. The American workplace fixation on time and the clock frustrates most new entrants who feel compressed and stressed by this micro yardstick. This mindset and way of life have produced an efficient and thriving society, though often at the expense of interpersonal connectedness and relationships, some argue.

Childrearing has an American tinge as well. In the American culture, parents speak to their children as adults and start early to imbue personal responsibility for their actions. Also, they encourage their children to question and always ask "why". They argue, this behavior prepares the mind for curiosity and the window to creativity and understanding. Babysitters are hired to cater to the children when the parents are away from home. It is mandatory for children from ages 6 to 16 to attend school. Newcomers quickly observe how the school system provides buses, computers, and subsidized meals for indigent students. In most cultures, the students and their parents are left to fend for themselves. Students are expected to ask questions in the classroom. Parents do not physically hit or inflict bodily harm on their children. Such an act is regarded as child abuse.

Only the child's parents may discipline him or her. They discipline their children by taking perks and privileges away. Matilda, who emigrated from South America, told her family's harrowing experience with the educational and justice system so many years ago. Her entire family comprising her husband and three children were legally admitted into the United

States under the Visa lottery program. It was tough for them to emigrate since the husband held a good and influential government position, while Matilda worked as a high school teacher. Economically, they fell within the middle-class ranks in their old country. The fear of the unknown in a new land almost derailed their decision to emigrate.

Nonetheless, they eventually migrated and settled in the country's Midwestern region. Both parents were quickly gainfully employed. Their oldest child Thomas was in his early teens, and as such, the teen hormones and youthful exuberance were emerging, with some sprinkle of rebellion. This behavior was alien to the then young amateur teenager's parents. They resorted to the disciplinary methods of the old country. This correction involved harsh talk and sometimes lashing with objects such as flip flops, or belts, or any adaptable item. Initially, Thomas stomached such seemingly draconian discipline until he started to share his experience with a couple of his new school friends.

One of his new friends reported Thomas's story to a schoolteacher. Thomas was invited for some probing and inquiries. Thomas was advised and encouraged to report any hitting or lashing to the school authorities. Just two days after the directive to Thomas, he received what he had termed the mother of all lashings for behavior he admitted was beyond the pale. Thomas agonized all night whether to tell the school authorities and thereby expose his parents' "normal" disciplinary actions in the old country but unacceptable practice in their new land. As a brilliant kid, Thomas had reasoned that their parents were still learning the values, culture, and way of life in their new nation and may deserve some break at this early stage of their assimilation. This reasoning went out of the window once he got to school, and his friends urged him to report the incident immediately. He yielded and reported it.

The assistant principal subsequently invited the parents for some talk. The talk did not go well. This breakdown was because, from Thomas's parents' perspective and worldview, parents can discipline their children any way they wanted as long as they inflicted no physical injury. They had stormed out of the vice principal's office in a rage that the school was trying to control how they disciplined their child. In telling this story, Matilda kept pointing to her stomach when she emphasized that it was her child that she carried for nine full months and delivered him with her husband's support. Now, someone else was attempting to dictate for them how to raise him. This story illustrates the collision of two well-intended cultures. Over the years, Matilda and her husband finally understood what took place with that episode. A confession-like statement reflected their initial naivete and lack of a deeper appreciation of American culture.

In some cultures, the discipline and character-building of the child are communal. This communal childrearing takes me back to my early formative years in Africa. A group of friends and I were experimenting on smoking cigarettes. We had pooled our meager resources to purchase a pack of the cheapest cigarette. We settled at an alley we thought was safe from the watchful eyes of adult neighbors. Little did we know that an area busybody mom observed our every move and activities from her upstairs window. As we started to puff away, she cracked her window open and shouted, "Okay, you, you, and you (pointing and bellowing out our names), I will report you all to your parents, and put out those cigarettes and throw them away, you all know better." We immediately put out our cigarettes and scrammed. That evening, each of our parents received a visit from this lady, who proudly reported her observations. All parents thanked her profusely. Such was the culture. No parent batted an eye or questioned this lady's motives, intentions, or rewards. The society and culture of communal childrearing were prevalent and embraced.

How about this one. Grace was twenty-one and was a junior in a local university in an African city. The city boasts of a regional airport among other edifices. Grace had flown to visit a sugar daddy in another town and had flown back with a commercial airline that evening. Seated among the passengers was an older man, popularly called Pa George. Grace told the story that she cringed when she saw Pa George, and her heart rate accelerated. Why? You would ask. You see, Pa George was a respected local icon and a good friend of Grace's family. Pa George was well aware of Grace's family's financial capabilities and would therefore surmise that Grace's family could not have afforded the flight ticket. What was more, Grace was supposed to be hunkered down at her campus studying tirelessly. What on earth was she doing in this other city, Pa George wondered. Grace was correct in her concern because soon after Pa George saw her, he walked up to her with a fatherly inquisitive tone. "Hello Grace, how are you?" and before Grace could muster an answer, he continued, "and what are you doing in this city instead of being on your university campus?" As Grace started to formulate an immediate cogent answer, Pa George followed with, "Do your parents know that you came here?" Grace responded, "Not really." At this point, the damage had been done. Pa George walked away in disappointment. This episode, expectedly, did not end there.

Upon arrival at their destination, Pa George was picked up by his chauffeur, whom he directed to drive straight to Grace's family house. There, he informed Grace's parents how he ran into Grace at the airport traveling from her university campus city. Grace's parents thanked Pa George for the information and well-meaning care and concern. This encounter, interactions, and tattling were proper and normal in that culture. Grace's parents did not view Pa George's action as intrusiveness in their family's affairs. Though Grace was twenty-one years old and away at a university, she expected Pa George to tell on her. She knew and grew up in this culture and value system that preached and practiced communal childrearing. She explained that she would have been surprised and interpreted it as uncaring if Pa George did not "report" her to her parents.

Another subtle but important aspect of the American culture rests with the eye. Eye contact, to be exact. In America, people look you squarely in the eye and shake your hand firmly. These behaviors represent the embodiment of directness and openness in interaction. This topic takes us to Asia and the experience recounted by Scott. Scott was an executive for an international conglomerate with offices and locations worldwide, including the Asian continent. In one of Scott's early company trips, he visited Japan. Scott relayed his naivete at those early exposures to Japanese culture when he misread some Japanese female managers' demeanor and behavior in their branch offices. According to Scott, these female managers would never keep a sustained gaze at the eye of whom they were addressing or talking with. They could laboriously make fleeting eye contact and immediately look away or down. Later, Scott understood that these gestures represented respect and not an inferiority complex. Scott observed that even when he trained his eyes on any female managers, they would look away or down. In the American context, a conscious effort to not look one in the eye could be interpreted as weakness, tepidness, and timidity. In job interviews, one could be eliminated from job candidacy by, among other reasons, not looking the interviewer in the eye.

When gifts are given or received in America, you are to open such presents or cards in front of people. A major part of the gifting experience lies in the gifts' instantaneous opening, allowing observers to ooh and aah. Imagine a new entrant accustomed to gifting experience in their old country where presents are quietly and privately opened, suddenly expected to defy that practice and openly rip their gift package. This expectation could be nerve-racking and discombobulating. But you have to manage the sleight of the hand you are dealt to integrate into your new land. It would be abnormal at funerals to make loud, sad, and harrowing sounds or noises. You are required to bottle your emotions even if you felt like bellowing a few loud ones. Wailing in some cultures is a part of the funeral rituals. An absence of loud and sometimes wild wailing is decried in some cultures. Professional wailers are indeed hired to add pomp and liveliness to an otherwise drab funeral in some cultures. These expert wailers train for their skills and art and mix this practice with entertainment. They are revered and respected for their unique talents and competency. In America, money is not given directly at funerals. Instead, flowers and plants are provided, though occasionally electronic solicitations for money are made via established apps such as GoFundMe, Venmo, and PayPal among others. Other times, donations are made to designated organizations in the name of the deceased.

You never go to someone's house without first seeking to know if it is convenient. Freedom, convenience, respect, and comfort drive this mindset. You cannot waltz into someone's abode, even if a relative and family member. If you are invited to a gathering, event, or party, you do not have your children in tow. Unless specifically indicated, bringing uninvited participants such as children or other friends would be considered rude and disrespectful.

In the United States, discrimination based on ethnicity and nation of origin, race, religion, sexual orientation, etc., is protected by law. The judicial system strives to dispense equal justice on these matters. It is generally successful but fails to rid society of discriminatory practices and practitioners. Any form of domestic violence is against the law. For example, it is illegal to hit anyone: a child, a parent, other relatives, a spouse, and even a pet.

America is a youth-centric culture. As a result, its citizenry works very hard to preserve their youthfulness, even if that entails some medical augmentation such as elective procedures and other medical interventions. A newcomer from South Korea expressed her shock and bewilderment when she first saw an American teenager with braces. She described it as wires and metals intricately woven in one's mouth. She had first observed this while standing with a relative at a store's checkout lane.

As the teenager spoke to someone who looked like the mother, the newcomer from South Korea's jaw dropped. She quickly turned to her relative and uttered in Korean, "Oh my God, what did I just see, some girl with wires and metals in her mouth? Is she able to chew those? Can she effectively eat with those? Poor girl, she may starve to death." It did not help that the teenager was thin and lanky. She was so strikingly affected by this spectacle that she was almost hyperventilating while speaking in her native tongue. The relative cautioned her to calm down, and that the procedure was standard here in America. As they walked out of the store, she trained her gaze on this girl exiting the store with her companion. So much so that she bumped into a parked car at the lot. Even after her relative explained that Americans care very much about beauty and looks and could be deemed obsessed with them, she could not wrap her head around what she saw. Would you blame her? After all, in her old country, crooked and messy teeth were not seen or regarded as a natural defect. No one would even bother to undergo an artificial and inhuman correction of your God-given attributes, she argued. She jokingly requested that her relative take her where she could have a close-up view of this incomprehensible "atrocity." The relative laughed her off.

Most older Americans prefer to live in retirement, independent, and assisted living homes rather than live with their children. The independent streak in citizens is a cradle-to-grave mentality. The American economic engine revolves around productivity and value brought to the workplace. When that value wanes or disappears, the once productive individual is relegated as an elderly to a retirement home. Some still find their relevance at these homes, while others feel unwanted, useless, and unneeded in society. The family dynamics and relationship become one that borders on semi-care. Visitations by family members are primarily out of obligation than showering of love. As the health of the elderly decline, caring and emotional attachment are rationed. Often this position is the most practical there could be. After all, would it make much sense sitting in the room of a parent who has severe Alzheimer's disease that does not recognize or acknowledge your presence?

Moreover, watching one's parents' health deteriorate could be heart-wrenching. Therefore, for survival and self-preservation, the degree of emotional lavish is calibrated.

Emigrating from a culture where the elderly is treated differently, such as living with family members, could initially find the American elderly experience different and challenging to understand. Nonetheless, culture is relative to its environment, and the treatment of the elderly in America is not an exception to this construct.

Death and dying are deeply rooted in culture. The view and treatment of these twin end-of-life occurrences can be dramatically divergent among cultures. Death is a family, communal, and religious event, though less so than before. Centuries ago, death and dying occurred at home. Therefore, relatives had direct personal experiences or tasks in the process. Today, medical interventions through hospitals, nursing homes, and hospice centers share some of the burdens. Some submit that institutional, professional, and technological mediation rob family members of the erstwhile hands-on responsibilities in death and dying. In some world cultures, death and dying remain intensely familial and private. It is the relatives who nurse the dying until death. It is the family that plans and arranges every aspect of the burial and final rites. There are no hospice centers or funeral homes.

Unlike some cultures that suppress death and dying and avoid its embrace and planning, America's approach to this subject is unique. For example, American healthcare providers are transparent in conveying possible end-of-life timeline to their patients, regardless of how insensitive that might appear. They are trained to tell it as it is and not build unrealistic expectations. Overly optimistic views and advice could ultimately be more harmful to the patient and the family, they submit. However, the American healthcare providers' candor represents a positive development in the death and dying sphere. For example, American culture is tagged of avoiding and abhorring death and dying. America is a youth-obsessed culture, and therefore, death is seen as distant and unconnected to us until it occurs.

Additionally, medical innovations have improved and prolonged life. So much so that death is viewed as a failure. Even when it happens, we use euphemisms such as "passed on," "passed away," "passed," "expired," "departed," "gave up the ghost," and "gone to meet their maker." These depict denial and skirting death. We generally allude to death and dying fleetingly in discussions. Consider that rather than being buried, we say, "laid to rest."

In 1989, a visiting relative from Africa and I visited a family whose head was diagnosed with stage 4 pancreatic cancer. The wife of the patient shared the doctors' treatment options. As we listened and waited anxiously for the recommended regimen, she sighed and stated that the doctors recommended no further medical intervention for a cure. Instead, they provided information and contact details for a hospice center in the city for palliative care. The doctors advised he had about three months to live, she concluded. Driving home, my visiting relative was beside himself that the doctors could be that callous and inhumane to tell a dying soul such awful news. You see, my relative's background and culture elicited such a reaction. In that culture, healthcare providers are trained to remain optimistic to the end. Such an end-of-life timeline would layer more worries and quicken death, they reason. Therefore, they hoard information, even with family members, often until the last moments. They contend that inevitability compels them to manage the death and dying process with caution, gradient, and grace. In these cultures, and even somewhat in America, the concept of death remains taboo. It is regarded as socially inappropriate and negative to discuss death openly. Death and dying are reserved only for when it becomes necessary.

The American counter-argument with the information-dripping approach is that patients could plan accordingly by providing the realistic status of illness and eventual death timeline. Moreover, America's attitude towards death is that once inescapable, all efforts must be exhausted to provide a dignified and less painful end where possible. The hospice movement ushered in gradual openness about death and dying in America. Hospices provide support to the dying and their relatives. There is a prevalence of death anxiety in America. Death anxiety is revealed in two areas; anxious about death and being anxious about the process of dying. Relatives and immediate family members concern themselves with death itself, while the elderly and presumed dying are preoccupied with the dying process. Will it be painful or without suffering, etc.? One could argue that this anxiety makes the hospice movement an acceptable and growing service.

The right to die debate continues to rage in America. The right to die emerged as a significant discourse in keeping an unconscious person alive with artificial and medical equipment. The debate expanded to include accepting the terminally ill's wishes, who are in extreme pain, to end their lives. This debate remains unsettled in America.

At funerals, we are encouraged to be stoic and seemingly emotionless. We are to hold it in and not shed tears. Such a display of emotions would be undignified and disrespectful. Sometimes, this pent-up feeling leads to unresolved grief that manifests in some fashion later. Before casually dismissing this practice, a quick contextual examination is warranted. Scholars posit that funerals perpetuate social order, help in grief processing and calibration, reinforce religious and faith beliefs, and offer an opportunity to demonstrate love and affection for the deceased. Whereas some cultures believe that the outpouring of emotions, such as crying at funerals, provides some healing for the soul, American funerals' perspective is different. In America, funerals are designed to offer the platform for organized and public sorrow, which could not be expressed thoroughly elsewhere. No wonder that the deceased family, relatives, and attendees' attitudes and behavior are rooted and manifested in anguish, loss, grief, and sadness. These negative emotions dictate a sorrowful disposition rather than a celebratory inclination at these events, regardless of the deceased age.

This book will not attempt to wade into or settle competing viewpoints on extreme candor or miserly information regarding death and the right to die positions. Instead, it aims to portray how a subject such as death and dying can be culturally based. One commonality among all cultures is that end-of-life ceremonies, such as funerals, burials, and memorial services, present a sense of closure after the loss. These seeming conclusions allow the survivors to express their final love and affection to the deceased.

Let us veer to some lighter side. Supersizing food and drinks is as American as apple pie. The sizes in America are large, larger, and largest. Bigger portions of food and beverages are celebrated and perhaps even glorified. We tend to go big or go home. Ask someone in Texas about this subject where bigger is always better! An interesting irony is while big portions of food are ordered, an equally tall order of diet drinks is placed. A mindset that perhaps, the healthier drink would mitigate the gluttony. A family from Greece that vacationed every year in the United States for twelve years straight relayed how the highlight of their yearly trips was betting on who amongst the family members would consume the largest portion of fries and soft drinks at a sitting. They complained that after their annual three-week vacations in the U.S., each gained at least seven pounds. They enjoyed the supersizing phenomenon, perhaps, because they knew it was a very temporary and once-in-a-year treat. In one of their visits, the oldest son insisted he could singularly devour a 20-ounce steak. The rest of the family took him to the task. He ate every bit of this mammoth steak but became sick as a dog after that. He recovered after some hours, thanks to over-the-counter medicine, Zantac. This food and drink indulgence usually ceased once they returned to Greece. The family patriarch commented that one of their children wrote a positive school essay on the merits of obtaining large portions of food and drinks at restaurants in America. Not surprisingly, he was twelve years old when he wrote the essay and had since recanted his supersizing endorsement for health and wellness reasons.

Still sticking with America's fascination and obsession with largeness and bigness, consider America's roads. They are very wide and multi-laned in most cities, and interconnecting highways are commonly referred to as freeways. A visiting European once joked that jumbo jets could easily land on most American highways. Until perhaps the seventies and eighties, when the oil instability and scarcity compelled America to build smaller cars, most cars were large and roomy. The metals and other raw materials sunk into one vehicle could have easily built two smaller automobiles. Some of the cars designed to seat five people could comfortably accommodate seven to eight people. These vehicles were road tanks that guzzled gas hungrily and ferociously, obtaining fewer road miles per gallon. These days, when one of these "dinosaur cars" drives down the road, the younger folks marvel at their ancientness and sheer wastage of materials and the natural resource of energy. Thank heavens that these vehicles no longer ply the roads of America. Observe, though, that these cars were just a manifestation of America's symbolism. They represented the big and bold identity of the United States. If not captured on the automobile, it is reflected on the roads' size, displayed by big houses, and showcased in the massive portions of food and drinks. They all depict America's love, obsession, and celebration of largeness.

Americans do not hold back their expressions, both angry and happy ones. We tend to let out heartfelt, open-mouthed guffaws uninhibited. Why not? We are free always to express ourselves. That represents a part of our freedom and an open society heritage. We flaunt this latitude through various avenues. Little wonder some cultures regard our boisterousness as rude and disrespectful. Too bad. We are exuberant, lively, and optimistic people.

Has anyone figured why we write our dates a certain way and most of the world expresses them another way? We write dates; Month-Day-Year, while most other societies represent them as; Day-Month-Year. Who was on first? Perhaps we went the other way as a sign of our difference and uniqueness? Who knows? If you think this is just a trivial item, ask newcomers who used to write the dates their old country's way. You will hear stories of the agony of initially writing checks with dates or signing agreements and documents that require inserting dates. It needed reverse dyslexia to get it right. During the initial years of my migration to the United States, I dreaded writing dates because it would cause me to stop on my tracks to get it right. After some years, it became second nature to writing it the American way. Whew!

A friend of mine at college told a story of a schoolmate neighbor who had emigrated from Europe and was invited by the native-born to watch a football game between the local team, the Seattle Seahawks and San Francisco Forty-Niners. The emigre had only been in the United States for three weeks. When my friend threw the invitation, the newcomer had thought the match was football, European or most of the world type. So, he was very excited to pass the time on a beautiful Sunday afternoon indulged in watching his favorite sport, football. Football in Europe and other continents is *the sport*. Life revolves around football in most of these countries. Relationships, including family, can be enhanced or ruined by this sport. Football fans around the globe have been labeled the most obsessive and rabid fans. Following, support, and loyalty to football teams and clubs are as ancient as the beginning of time. Little did the newcomer know that it was football, the American version. He hurriedly went across the hallway at the designated time to my friend's apartment. On his part, my friend did not know that football to Europeans meant soccer to Americans. He was as ignorant as his guest. Needless to say, the afternoon watch party did not go quite as planned. The invited neighbor could not understand the American football game's rules and format, much more enjoying it. He was a good sport, though, and they made the best of the situation.

Jeans, yes, jeans. It represents one of the classic American cultural footprints. Jeans is as American as there is. This iconic national symbol comes in different tints, styles, and fits. They include but are not limited to bell-bottom jeans, skinny jeans, baggy jeans, stone-washed jeans, stretch jeans, relaxed jeans, etc. Jeans signify America in its ruggedness, toughness, casualness, simplicity yet sophisticatedness, and universality.

Little wonder it can be deemed the "national uniform." It is worn by everyone, from the very elite, the wealthy, the serious-minded, the middle-class, and the poor.

Of course, not all cultures cherish and embrace this article of clothing like Americans do. Some cultures frown at its perceived extreme casualness and non-elegant appeal. The newcomer would quickly observe that jeans represent the most beloved wear in the United States, bar none. One of my relations, who had also emigrated to the United States at approximately the same time that I did, found it very difficult to integrate into the so-called jeans culture. He had viewed jeans as just clothing for the rough and tumble cowboy or cowgirl. In his mind, jeans were for farmers and related professions. It was only meant to be worn at the home and surroundings. So, for several years he adamantly refused to be caught up in the jean's euphoria. He would not dress down, he would state. He did not own any pair, such that when his neighborhood leaders organized a jeans attire-only picnic, he did not attend. How is the fairing with jeans these days, you may ask? Allow me to report that he dons jeans now more than most people I know. If there were a jeans pajama that one could sleep in, he would buy one. A few years ago, I teased him in the presence of his twenty-something-year-old children about his initial repulsiveness with jeans. He cracked a wry smile and chucked it to his stubbornness and inertia. You see, what my relation is dismissing as bullheadedness is larger than such a narrow attribution. He experienced some tension between the values, groundings, and tenets of his identity and his new land's realities at that period. He could not flip his identity overnight and may never veer from who he was as a person. Social scientists tell us that we could play at the edges of our identity but never completely alter who we are. My relation eventually embraced jeans and what it represents but remains true to his core self.

The United States, as the 3rd largest country in the world, is blessed with enormous land and natural resources. The vastness and various topographical richness are to behold. When white settlers entered America's shores, they were struck by what they perceived as unaltered nature. This observation left an impression as they ventured inland. Writers like Ralph Waldo Emerson and Henry David Thoreau celebrated nature as a spark behind the American character. Some argue that the wilderness produced the American identity, and the abundance of natural resources brought richness. The "frontiers" were embraced and romanticized. Art and paintings captured this appreciation. America successfully combined nature and civilization in its culture. Historian Fredrick Jackson Turner submitted that the frontier line separating civilization from the wilderness is "the most rapid and effective Americanization on the continent."

When one emigrates from a geographically smaller country, the American landscape's enormity becomes fascinating and magnificent. Often newcomers from continents and regions where a three to four hours' drive places one in another country marvel at the vast distances between American locations. Sometimes, one can drive for over eight hours in one state. Ask California or Texas residents, or perhaps the "Big Sky" Montana, and "The Last Frontier" State Alaska. What is enthralling is the various vegetation, flora and fauna, landscape including mountains, and climate and weather patterns throughout the United States.

A Swedish family who vacationed in the United States in the nineties embarked on a cross-country trip from the East to the West Coast. While traveling, they made some

Southern and Northern regional incursions. After their long and exhaustive experiment, they swore not to ever engage in such adventure. It was not because they did not enjoy their trip. It was because they had grossly underrated the vastness of America, geographically. When someone questioned if they had thoroughly planned their journey, they retorted that one could plan all you want on paper, but making the trip was different. They did not express their answer from a negative stance. Their observation was from an admirable and awestruck perspective. They remain strong cheerleaders that nowhere on earth (where they had visited and they have been to numerous continents and countries) had equaled the United States in terms of enormity and richness of terrain.

"Only someone else can scratch
your back thoroughly."
-African Proverb

CHAPTER 12

THE WORKPLACE

Work is central in the American culture. Work is treated as morally right and nonwork, an unacceptable proposition. Therefore, the United States is a country of hard workers. Most in America define themselves by their occupation and trade. An individual's profession contributes immensely to their sense of self, purpose in life, and general disposition towards their being. There is a strong tendency to live to work instead of working to live. Emigrating to the United States dictates that one puts on the proverbial running shoes upon arrival. To make headway, you must pay your dues by working hard.

Let us examine workers' peculiarities and the United States workplace to portray the newcomer's landscape. The work mentality demonstrates putting in long hours during the day. Due to the proliferation of technology and digital devices, additional tasks are performed at home. According to the Pew Research Center, 77 percent of Americans work more than 40 hours per week, and the American workweek averages 47 hours. Lately, the technology and digital incursions in workers' personal lives are more prevalent, prompting the coinage of the term "electronic leash" by companies of their employees. This referenced hold is usually voluntary as employees always seek to be connected to their work and organization. Breaks and lunch periods are not celebrated like many other countries' work cultures. In America, lunch is wolfed down in minutes at the worker's desk, and focus reverted to the task at hand most of the time.

Elsa, who emigrated from Sweden, initially described her impression of the American workplace as oppressive and unreasonably demanding. When she arrived in the United States, her first job was at a bank, first as a teller, then a personal banker. Elsa had worked at a major bank in Sweden and therefore was familiar with essential banking responsibilities and rituals. What she did not foresee before her arrival was the greater demands placed on a comparable position in her old country. Elsa quickly learned that in the American workplace, relaxed and hourly lunch was frowned upon. She learned this the hard way. In her first formal evaluation, her boss dinged her for taking excessive time for lunch. Naively, Elsa argued that she never spent more than the allotted time for lunch. Her boss demurred by stating that she was sometimes expected to skip lunch or take a quick one depending on workload and bank customer traffic. That was a huge "learning moment" for Elsa. Welcome to the American workplace! All the while, Elsa thought she was fine sticking to the company-provided lunch period. Her boss further explained that she needed to gauge the urgency of any moment to determine how she responded to her lunch outings. This mentality was a shock to Elsa as she looked forward to her daily lunchtime. Elsa reasoned that inadequate lunchtime was depriving her of the workday's highlight. Welcome to America!! Elsa eventually left the bank but not because of the lunch matter and learned an important lesson in her assimilation into the American workplace culture.

Please indulge me in another workplace cultural shock experienced by Pierre. Pierre recently emigrated to the United States from France. He arrived under the Intracompany Transferee Visa program for special skills and competencies. He was excited about his reassignment and transfer. So, he came in and plunged right into his responsibilities, which was in the cybersecurity sector. Before leaving France, he was practicing the national law that detached employees from their employers after work hours. France enacted this law to address the then-emerging employee mental health and well-being issue. The clamor for this law emanated from employees' after-hours work pressures and expectations. This law explicitly absorbs the employee of not reading, reacting, or responding to an after-hours email or other electronic communication forms, including phone calls. Pierre had lived by this practice and behavior and sang praises to its effectiveness in providing needed decompression for employees. Therefore, it had almost become second nature for Pierre not to check or respond to work-related electronic messages after office hours. Pierre essentially switched work off daily once he left the office. He claimed that his wife commented about seven months into observing this law that she had noticed a more relaxed and pleasant husband. According to Pierre, she stated, "I have gotten my husband back again." This exhilaration was because he no longer brought work home through electronic connectivity.

Then, Pierre and the family emigrated to the United States. At first, the gradual creep into the now forgotten routine of electronic connectivity to work 24/7 was stealthy. The rude awakening occurred when, one day, Pierre was berated by his boss for not responding to what the boss had deemed an urgent matter after being sent an email at 8 pm the previous night. It was not enough that Pierre had responded to the email at 7:49 am the following day as soon as he got to the office. This experience evoked some flashback for Pierre on the pre-right to disconnect law enactment and implementation. Memories of the heightened stress levels and personal time incursions came flooding back. What made matters worse for Pierre was that he was no longer in France. He was now in the United States, where there are no laws allowing employees to disconnect from their employer legitimately and legally. Pierre ultimately had to readjust to this new reality of his new land.

The American work reward system is individualistic. Organizations evaluate and hire candidates based on what value they can bring and their ability to work as a part of a team. But make no mistake about it. It is all about individual accomplishments in the long run. It is a results-oriented focus. This workplace feature fuels competition either directly or inadvertently. In chapter 11, we discussed time and punctuality as a way of life in the United States and highlighting their importance and premier status in the workplace. If time and punctuality were held normally in the greater society, at the workplace, it is on steroids. Punctuality portrays the seriousness of purpose, steadiness of character, high level of trustworthiness, and respect for your duties and organization.

Most companies and organizations in the United States operate on the modern organizational leadership theory and empowerment approach. This style differs from a command-and-control structure. As an employee, your opinion, input, and ideas are sought in improving procedures, processes, and deliveries of products and services. The degree of employee input may vary among companies and organizations, but the management and leadership stance espouse teamwork and collaboration. The ability to effectively oscillate between solo and teamwork is celebrated and rewarded.

Change and the embrace of change are permanent fixtures in the American workplace. The change could be reorganizations, tasks consolidations, realignments, personnel layoffs, minor assembly line tweaks, or policy updates. Nonetheless, constant change is an integral part of the workplace reality. In the United States, organizational structures, methods of performing tasks, policies, and procedures are ever evolving. The only constant feature in this environment remains change. Company acquisitions, consolidations, reorganizations, and re-engineering occur at a breakneck pace. Some have argued that the incessant change causes instability and disruption of the economic landscape.

In contrast, others maintain that the American economic model and its nimbleness and adaptability guarantee a robust and thriving economic system. Innovation and organizational agility are greatly valued in the United States. Some cultures and economic systems cherish and celebrate the status quo's stability and maintenance. In those places, change is frowned at, and constant change is detested. Organizations that demonstrate stability and steadiness rather than continuous engagement in structural, operational, and procedural changes attract job seekers. Organizations would boast of not experiencing, implementing, or introducing change for years on end. Such is not America. Expect and get used to constant change brought about by internal and external factors, including global phenomena.

A highly recruited executive from South Africa was strongly encouraged to attend seminars and conferences on change and change management. His boss, the president of his division, espoused adeptness at instituting and shepherding change within the enterprise. This new executive became confused and disillusioned by the strong emphasis on change. Change is a necessary part of the business environment but not a concept that must be "worshipped" as it seemed to him at the time. He believed that the perceived lopsided preoccupation with change and change management dwarfed other vital aspects of managing and leading an organization. He vigorously argued against induced change brought about by non-prevailing events and circumstances. He submitted that such an artificially manufactured change process disrupts otherwise well-functioning systems. He further contended that constant jerks on an organization and its employees breed a demoralized workforce and employees' overall fatigue. His boss, who was native-born and ivy-league educated and trained here in the country, saw it differently. His boss strongly believed in anticipating the future, looking "around the corner" as he would preacherly posit and injecting change to inoculate against the future and get ahead of the competition and peer organizations. You could not blame him. After all, he successfully grew the company from a regional powerhouse to a notable global player in their sector and niche of agribusiness. The boss maintained that with the increased proliferation of technology coupled with the globalization of markets and industries, speed and agility must take centerstage in organizations. As a result, he aimed at always structuring the organization to conceive, implement, and absorb change efficiently, effectively, and at jet speed. This posture and strategic bent served the company well over the years.

For the executive recruit from South Africa, the pace of change was disconcerting and too disruptive for his liking. Eventually, he left the organization. In reflecting on the recruited executive, the division president observed that the newcomer could not break the cultural, mental block regarding change and its practical use and application in leading an organization. Successful and thriving organizations in America embrace or even induce and manage the change process effectively and adequately. It is often said that the only thing constant is change, and this adage squarely applies to the American workplace.

The American workplace culture involves meetings that cultivate and nurture brainstorming, consensus where applicable, collaboration, and teamwork. These gatherings focus on information and data analysis, planning, monitoring, and reviewing items and projects. Wholesome participation of all is usually solicited and sought. Meetings represent the platform and avenue for formulating solutions to problems. Though, at times, meeting purposes and goals are misguided and unproductive. Nonetheless, they remain relevant in the American workplace context.

Another component of the American workplace lies in the encouragement and practice of friendliness. Organizations invest substantial resources to establish, foster and promote congenial work environments. Sometimes, the focused desire to establish a friendly atmosphere downplays criticism and conveys a superficial commitment that stifles robust and healthy debates and dialogues. In other words, to avoid confrontation and crucial conversations, direct and corrective discussions are circumvented and sacrificed for congeniality. Attempts at foisting congeniality at the expense of challenging and productive discourses could be counterintuitive.

Understanding that you are no longer in your old country rife with its workplace culture helps usher accelerated workplace assimilation. Observe that if you emigrated from a country whose workweek runs Sunday through Thursday, you kiss that goodbye. If you migrated from a country where you were once protected by a "Right to Disconnect" law, which allows you to legally not be responsible for responding to emails that come in after-hours, you wave that bye-bye. Yes, this law is alive and fully functioning in France. If the norm and expectation in your old country allow you to waltz into a scheduled meeting fifteen minutes late, your paradigm and practice must change. Meetings start promptly, or let us say, almost promptly but not thirty minutes late. It would be deemed disrespectful and rude to show up grossly late to a meeting. If you came from a society where organized communal fitness is practiced at the workplace, you are now on your own. You choose when to go to the gymnasium and do your thing. It is the "freedom of choice 101" in your new land.

Suppose you are from a society where a specific time is allotted for coffee to enable all workers to congregate for conversations and camaraderie. In that case, you can obtain and drink your coffee whenever you want, based on your organization's rules. There is no collective handholding and "kumbaya" singing during coffee breaks. Suppose you emigrated from a land where discussions and positions on topics and ideas are posited abstractly through stories, allegories, and insinuations. In that case, you must become used to directness and candor in expressions. Thought processes and perspectives are conveyed explicitly and unabashedly. If you emigrated from a country where after welcoming a child, each parent is given three months parental vacation with 80 percent of pay while on leave, sorry, not here in the United States. If you are from a society where you can plant a "mock kiss" on your coworkers' right cheek as a sign of welcome, respect, and affection, please do not attempt that gesture or anything remotely close to that in your new American workplace. It is unwelcome and prohibited by rules and the law, as it may constitute sexual inappropriateness and harassment.

Lena emigrated to the United States seven years ago from Poland and had secured a job as a supervisor in a hotel organization. She was very hardworking and diligent in her responsibilities. Lena's staff and supervised employees were majority males. Almost daily, Lena would adulate some of her male workers on their outfits with words such as "You look stunning in that outfit today," "That shirt looks so great on you," "Your pants fit your body contours so well." These were mere positive, complimentary words to boost the employees' spirits at their morning encounters with Lena. However, in Lena's new land, America, these seemingly harmless compliments represented potential sexual and workplace harassment issues. No sooner than two months into her new job, Lena was summoned by the organization's Human Resource manager to admonish her on utterances to male employees. The manager had received two complaints from employees regarding Lena's "compliments." At first, Lena was astonished as she explained to the manager that she freely doled such compliments to her supervised employees, who heartily consumed them in her old country. The manager immediately reminded her that she was no longer in her old country. The manager calmly explained to Lena the implications associated with these compliments. Lena was subsequently enrolled in a company-sponsored sexual and workplace harassment training.

Some years ago, when I was privileged to work for an international company that operated in eighty-eight countries, it was fascinating and a great learning experience to travel to some of these countries to participate in business meetings.

You could easily spot the Americans at lunchtime. This starkness was because while our host executives were merely getting ready to engage in a relaxed and unrushed lunch, the Americans would wolf down their lunches as if they were on a timer and sit on the table for the rest of the lunch period looking like bored teenagers among parents. Americans were generally wired to rush through their lunches and get back to the task at hand. I remember one of my American colleagues complaining that he could never get used to drinking wine during lunch. In some of the host country's daily meetings, fine wine was served along with elaborate cuisine at lunchtime. My colleague rationalized that his brain and body were never ready to absorb and enjoy exquisite lunch and fine wine until later in the evening or at night. He would grudge over the serving of wine at lunchtime. He judged the host country's workplace practices and etiquette from his lens. He forgot to read the chapter on cultural relativism in the big book of culture.

How about this one that never fails? Whenever a problem is teed up for discussions and deliberations at these international settings and meetings, we, Americans, always became impatient at long drawn out discourses. At least, that was how we perceived those dialogues and bantering. After just a short period of discussions, we were ready to offer solutions and a panacea, even when others from other cultures are just beginning to dissect the topic. Do you blame us? We are wired and trained to "fix it" and "make it happen" and quickly too. Therefore, once we believe we understand the problem, we rush off to the solution stage, no matter how skeletal. Maybe that is precisely why social scientists advance that we tend to be linear in our thought process, while some cultures are circular and even circuitous. None is better than the other. They represent two different ways of approaching a matter.

Interestingly, Americans' impatience at these meetings was equally met with frustration by non-American participants, who could not stomach our intransigence. I recall one of our host colleagues whom I played recreational lawn tennis with during those international meetings complaining that we, Americans, always seemed to want to demonstrate superior knowledge and expertise on matters. I attempted to explain to him that this behavior, which to a large extent is attributable to culture, education, and training, usually comes off unintentionally. I indicated that such conduct does not stand out in the American context and milieu and, therefore, not noticeable. He shrugged his shoulders, which I interpreted as "whatever." He did not buy my explanation. I left it alone, and we played our competitive tennis game as always.

Ever heard of potluck, especially in the workplace? Potluck refers to a gathering, usually at work, where individuals often bring homemade dishes or food to be shared. This event has become very commonplace in the American workspace. This practice's genesis is unknown or well-documented, but some attribute this practice to the native Indians centuries ago. Regardless of its origin, it has become a feature in the workplace. So much so, that for example, in the State of Arizona in 2016, the State lower legislative chamber considered and passed a bill reforming State law on potlucks and noncommercial social events. The legislators advanced the bill to regulate circumstances where people can serve food to protect against foodborne illnesses. Some have argued against workplace potlucks for what they term an extra burden on tired employees to prepare for such an event. In addition, they claim, though a voluntary proposition in organizations, forces an uninterested employee to participate to avoid not branded a non-team player.

Conversely, some have posited that potluck events promote camaraderie and team building in the workplace. They go as far as maintaining that potluck enhances connectivity with fellow workers over food. The cultural diversity benefit is not lost, they conclude. By employees from diverse backgrounds sharing their ethnic foods and cuisines, the atmosphere and environment are enriched and ultimately enlightens all workers. So, dear newcomer, welcome to the workplace rife with potlucks and similar events. It is a part of the American workplace terrain and tradition. Learn to cherish, embrace, and engage in this American workplace ritual.

In the United States, work commutes have increasingly become longer. According to the U.S. Census Bureau, the average daily commute is 53 minutes to and from work. This phenomenon is attributed to suburban sprawl and growth beyond the city centers. Realization of the American Dream entails homeownership. Most thriving individuals, couples, and families find affordable housing in the metropolis's outer limits. Consequently, it creates longer commutes to the workplace. Therefore, be it train, bus, or personal vehicle, chances are the newcomer may be subjected to longer commutes than previously experienced. However, with remote working gaining more widespread traction, especially against the 2020 COVID-19 pandemic's backdrop, weekly commute times are expected to be reduced accordingly.

In the preceding chapter eleven, we listed the attention to time and time management as one of the American cultural ecosystem's beacons. If time and the clock were your friends before emigrating to the United States, you most likely might develop a different, often negative, or at least cynical view of time and clock once in America. This pessimism is because time and the clock reign supreme in the American workplace. You are measured by and with time, or perhaps time rules over you as an employee. Your remuneration is dictated by time, especially as an hourly employee. Even with exempt employees on a salary, time directs activities and tasks. A newcomer from a culture where time is treated lackadaisically experiences the unbearable and suffocating emphasis on time. Imagine Timothy who was not only perplexed but annoyed that he was not paid his full wages because he had been ten and fourteen minutes late at two different times in one week. He had conjured up reasons for his minutes of lateness that fell on deaf ears with his supervisor. His supervisor rightly reasoned it would be unfair to other employees who were equally docked in the past when they were late for the same amount of time. This episode marked Timothy's rude awakening to the rule of time and the clock in the American workplace. He was never used to strict enforcement and modulation of time in his country's jobs. The clock did not rule there. Now yes, welcome to the American workplace, Timothy! A complete adjustment to time, minutes, hours, and days is warranted for alignment with the American workplace psyche.

The essence of highlighting some of these behaviors, norms, policies, and laws is to illustrate cultural relativism and emphasize that now, you are in your new land and must quickly learn the norms, expectations, appropriate behaviors, and laws that apply for faster adaptability and practice. You are not asked to rid yourself of your essential cultural background because that would be impossible and unwarranted. You are who you are due to your cultural foundations, among other influencers. You can adapt to the realities of your new land while holding tight to the underpinnings of your makeup. For example, you would not change your food and cuisine to adapt, and you would not change your religion to belong. You would not change your friendly disposition, if you were wired that way, to assimilate into your new land. You would not change your ancestry or nation of origin to acculturate. Some scholars have submitted that biculturalism leads to practitioners' better overall well-being. Social scientist and author John Berry described biculturalism as comfort and proficiency with both heritage culture and new land's culture. He posited an individual could surgically practice the application of both cultural flavors as the occasion calls. Such contortions could turn one into a human pretzel, in my opinion. Nonetheless, individuals who possess the needed cultural dexterity and discipline could be successful at such transitions.

That brings to mind one of my erstwhile executive colleague's agility and fluidity in transitioning from a mainstream posture to his Spanish heritage. I used to observe with admiration how this ivy-league-university-educated leader would swing from engaging in an eloquent English language discourse to a Latino-accented dialect and lingo and back to dominant English in a blink of an eye. Incredible! This skill would be more pronounced during the annual national meetings and conferences involving participants from many countries worldwide. He would hold informal gatherings during breaks and lunch periods as most South American and Spanish-speaking attendees flock to him for casual conversations. He would immediately switch to his rudimentary Spanish accent with all the gesticulations to boot. It was usually a sight to behold. I imagined that he attracted the participants due to sharing a common heritage and culture that they perceived. Being a division president, they most likely felt one of their own at such a high level and were probably proud of him. What was fascinating about this was that my colleague was born in the United States. He was a third-generation immigrant. He later explained that the family's Spanish heritage was not only instilled in all within the extended family network but celebrated. According to him, every member of his nuclear and extended family practiced biculturalism. He concluded by submitting that biculturalism allowed him to experience and enjoy both worlds comfortably. He would not change it for anything, he asserted.

"We rise by lifting others."
-Robert Ingersoll

CHAPTER 13

THE DIFFERENT CHARACTERS IN THE WORKPLACE

In this chapter, some of the workplace characters and personalities that the newcomer is bound to encounter are discussed. Again, actual names are not used, but the scenarios and depictions are real. These examples are not all inclusive but represent a bird's eye view of activities, interactions, and occurrences between individuals within the workplace.

Gary had been with the organization for over twenty-two years and worked hard every bit of those 22 years. He earned it. However, Gary could not contain his intolerance for immigrants, especially the newer ones. Gary wholeheartedly subscribes to the famed life coach, author, and educator Steven Covey's description of the "scarcity mentality." Scarcity Mindset believes that the pie is limited. It reflects the notion that there will never be enough to go around, leading to feelings of fear, stress, anxiety, and in some cases, aggressive behavior. It is an outward exhibition of lower self-worth and insecurity. It shapes our choices and our actions.

Attributes of scarcity mindset include believing that situations are permanent and fixed. No status is ever permanent! Someone with an abundant mindset, the opposite of a scarcity predisposition, sees life as dynamic and moldable. We can shape it to our desires. A scarcity mentality uses thoughts and words that mirror our mindset, usually negative in manner and nature. Our prism of viewing the world becomes our character and our makeup. We become envious, jealous, and downright intolerant of others. Psychologists tell us that one with a scarcity mentality generally fails to appreciate the blessings they have in their lives. They tend to focus disproportionately on the circumstances of others than their own. They are also not generous. They give less of themselves. This stinginess does not apply only to money. It includes kind words, smiles, time, and grace towards other people. In Gary's viewpoint, newcomers to the United States deprive the entrenched citizens of some opportunities that belong to them, the established citizenry. Is one surprised by Gary's behavior and antics, having possessed a scarcity mentality? Scholars submit it is all in one's mind.

For example, Gary may have felt that he had not gotten ahead because newcomers snatched away the predestined jobs and promotions he would have had. You see how groundless and perhaps illogical that mindset could be. Gary was on a project team consisting of six employees, and two of the members were newcomers from Australia and Asia. Gary stuck it out and grudgingly worked with team members. All the while, these newcomers had figured Gary's attitude and interactions with them were intolerant. They, too, found a way to be functional with Gary. Bigotry because of one's nation of origin happens. It is real. But you must deal with the hand you are dealt.

On a scheduled project team meeting, Ted from Australia and Gary arrived early at the meeting room. A casual chat ensued, and Gary let out that he was engrossed in a dilemma in his marriage relationship, and things were spiraling speedily out of control. As Ted started to chime in, another team member walked in, and they suspended their developing conversation and agreed to continue after the project meeting. Shortly after the meeting, they continued by Ted asking some intimate questions about Gary and his wife's relationship. You see, Ted moonlighted in Australia as a marriage counselor, in addition to his regular job there. Ted offered to visit Gary and his wife at home, and Gary accepted but asked that he clear that with his wife. Two days later, Gary informed Ted that it was a go, and they set the day for the visit. Alas, a breakthrough! Gary had begun to view Ted not just from the new immigrant perspective. Ted was a well-rounded person full of life, who happened to have emigrated to the United States. Ted made the visit as planned and with a couple more visits to Gary and his wife's home, he mediated and counseled them on their quarrel and misunderstandings. Ted became a marriage-saver, at least for that moment. Ted and Gary's families became close, such that they sometimes held dinner outings in area restaurants. Gary, whose attitude and perspective on newcomers was jaundiced, became more enlightened by this experience. That newcomers are just regular people like him trying to better their lives, registered strongly on Gary.

How about Matthew who would not mentally accept his newly emigrated and minted boss. Matthew confessed to doing what borders on bizarre to avoid interacting with his boss. Rarely would Matthew extend common courtesy to his boss in the morning by saying the typical "good morning." Instead, he would walk a long hallway to avoid his boss. It did not end there. Matthew would bad-mouth his boss instantaneously, usually saying derogatory and uncomplimentary things. I did not make this up. Matthew confessed to these stories. What made this behavior by Matthew disturbing and hurtful was that it was based solely on his boss' nation of origin and birth. It is real. It happens. Matthew eventually resolved his negative feelings and prejudices and made-up with his boss. He credited his catharsis and conversion to his religion and faith and has remained vigilant to avoid any relapse, he maintained.

Dr. Simon Knowles, popularly known and called Dr. Simon, was an accomplished management professor and author who occasionally engaged in public speaking. In 1992, he spoke at a management conference in Las Vegas. Approximately 400 people from different organizations attended. Dr. Simon, who had emigrated to the United States eight years earlier, was an eloquent speaker and presenter whose deliveries were usually unique, informative, and entertaining. He usually wowed his audience. He told the story of an incident that occurred after his speech and subsequent break during the conference. As people milled around, an attendee came up to him and commended him on his presentation, noting that it was informative and thought-provoking, and nervously posed a question to Dr. Simon. You see, Dr. Simon's presentation topic was "Globalization and the Future Workplace Culture." Some scholarly and scientific data, stances, and projections shook this gentleman's belief system and worldview. This confoundment was evident by his question, which was blurted as follows, "Your premise during the presentation was that the world economy would become blended and morph as one intertwined giant global economy …. And if so, where would the sovereignty of nations be, and how could that vision ever be possible?" Dr. Simon's eyes lit up as he recounted, and he licked his chop because he saw an opportunity to educate someone. Therefore, he spent the next 10 to 15 minutes imparting his scholarly knowledge on this curious conferee.

Dr. Simon stated that as their conversation progressed, he concluded that the gentleman's main issue was what the man had referred to as the mass influx of immigrants to the United States and how such inflows are diluting the talent pool and dulling the creativity and exceptionalism of America. Poor "Gentleman." Firstly, he was standing and speaking to an acclaimed professor who just emigrated to the United States and whose speaking his company paid thousands of dollars for him to listen to. Poor "Gentleman." The train of globalization had long left the station while still buying a ticket for the ride. Poor "Gentleman." Little did he know or realize that no one individual, no one nation, or even a group of nations could stop the globalization phenomenon and dynamic. It had already taken off and justifiably so.

Technology and digital advances were bound to make the world smaller, and they made the globe more intimate and compact. Such reach and interconnectedness opened up new markets even for American companies. Furthermore, it ushered in more efficiency in products and services manufacturing and delivery. Components of equipment or machinery are produced at various parts of the world with lower production costs and sent to the central assembly location or nation for final assembly. In turn, this process results in the final consumer paying a relatively lower price for the equipment, appliance, or machinery. Cases in point include computers, TVs, cell phones, and various household appliances. Poor "Gentleman." He was still stuck in the so-called scarcity mentality discussed earlier. Unfortunately, many may seem not to have elevated their understanding of globalization and associated cultural implications. Consider that most conglomerate organizations operate from numerous locations around the globe. Cross-border interactions, teams, personnel exchanges are inevitable and imperative. Understanding and appreciating the various cultural environments one must function in represent a required competency in today's workplace.

I cannot pass up some notable and perhaps legendary global snafus. Before globalization was chic and mainstream, it was alleged a car manufacturer produced a new model car branded "Nova." The name was acceptable in the nation of manufacture. These cars were then shipped to be sold in another country, and sales were meager. The manufacturer raised questions and discovered that the car's name "Nova" meant "No Go" in that country. Who would buy a vehicle that signifies "No Go"? Needless to say, it bombed. How about when the "Pinto" car was to be sold in Brazil only to find that the name was slang for "tiny male genitals." Ouch! Let us stay with cars for another one. Mercedes-Benz entered the Chinese market under the name "Bensi," which meant "rush to die." Oops!

Let us examine a few more. The chicken giant Kentucky Fried Chicken (KFC) sorely found out that its slogan "Finger lickin' good," translated in Chinese, meant "eat your fingers off." Gerber, which manufactures and sells baby food, introduced their products in Africa with a baby's portrait on the label. The problem was that labels are used to show the inside content of a package because of illiteracy. Double Ouch! Clairol launched a curly iron called Mist Stick in Germany. The problem was "mist" in Germany is slang for manure. Finally, when it translated its slogan, "Turn it loose" into

Spanish, the beer maker Coors found that it meant having diarrhea in colloquial terms.

The point in all these examples is that as members of the world community, we all must be culturally sensitive and demonstrate a heightened level of awareness, especially as the world is shrinking, so to say.

Back to people. We have got to share Helen's story to perhaps learn from it. Helen worked for a hotel chain in Chicago as a regional sales director. She worked her way up, first starting as a sales representative. Helen was well-respected in the hotel industry as a go-getter with a remarkable instinct. She had eight managers that reported directly to her. Of these managers was Paul, born and raised in a community about two hours away from downtown Chicago. He attended college in a nearby town. His exposure and worldview were somewhat limited. Of all of Helen's direct reports, Paul always was sarcastic and sometimes blunt about Helen's background. You see, Helen emigrated from Africa a couple of decades ago. According to Helen, some coworkers had intimated Paul's water-cooler negative comments, mostly centered on her nation of origin. He would comment and deride Helen's slightly different accent. It seemed that Paul could not stomach having Helen as a boss. Paul took an interest in asking Helen, in a "friendly" way, any negative news emanating from Helen's old country. By the way, Helen had become a bona fide American citizen. Helen always took the high road and avoided any entanglement with Paul on perceived intolerance matters. Until one midday in October...

Helen was having coffee in the breakroom when Paul walked in to obtain some brewed coffee. Then, Paul opened his mouth and stupidly asked Helen .... "Hey Helen, when would you be going back to your country, or do you like it so much here that you will forever stay here?" Helen's blood started to boil even before Paul was through with his inquiry. Being the consummate professional who would not go off her skids, Helen waved Paul to a nearby table for some heart-to-heart conversation. Helen began that she was a bona fide citizen of the United States just as he was, and this country belongs to all of her citizens, and equally. Helen felt this was the moment she had patiently waited for months and even years. Because, somehow, Paul had always felt and acted as though she was a second-class citizen since she was not born in this country. Paul expectedly stared at Helen hopelessly and haplessly, as if to suggest, so what? To Helen, it did not matter if Paul got it or not. She was satisfied that Paul provided that opportunity to get at what she perceived as the crux of Paul's intolerance. It was real. It happens. You must deal with the hand dealt you appropriately and adequately. Many Pauls exist. You must learn how to navigate the sometimes challenging and treacherous journey of the newcomer amidst these characters that dot your path.

A seemingly emerging phenomenon in the workplace has been loosely described as "immigrant on immigrant conflict." This tension manifests by a dysfunctional and non-congenial dynamic and the relationship between two newcomers in the workplace. This behavior goes beyond the normal emotions in any personal relationship. It is often rooted in competition and the desire to outshine each other. Generally and subconsciously, newcomers are lumped into one category. These newcomers know this too well. So, they vie for recognition, rewards, and possibly a promotion within their sphere. The result becomes the creation of an antagonistic air between the competing parties. Look closely in your work environment, and you may detect this usually destructive but subtle battle that often flies under the radar. Watch for sarcastic comments and put-downs disguised as matter-of-fact statements, avoidance of participation on the same team at work, or just plain tattling on the other party. Oh yes, it happens!

Some keywords, phrases, and comments to observe include but are not limited to; "Things are not as dire in his native country as he purports", "She is just clamoring for some sympathy and pity", "Make sure you properly check out her credentials, especially those obtained in her native country. They are usually not worth more than the paper they are printed on", "Beware of the skills and competencies projected. Beneath the splash, nothing substantive exists", and lastly,"He claims to belong to a relatively wealthy family in his native country, and that is not the case". The rivalry and competition produce these toxic and unfriendly swipes, which the recipient must filter. And this is precisely the message here. Recognize that the undercutting statements are fired to diminish one's "opponent" in this workplace cold war. Bring and utilize your barometer to evaluate and assess these situations to reach meaningful and objective conclusions.

Okay, Okay, where are the good stories? Where are the positive, feel-good people encounters? That is not the point here. Hundreds, if not thousands, of positive stories, abound. The American citizenry is known and celebrated the world over for its benevolence, good-heartedness, tremendous spirit, and overflowing kindness. Newcomers experience wonderful neighborliness, community embrace, and good-Samaritan-like gestures and treatment by entrenched citizens. The purpose of examining and exposing these seemingly non-complimentary episodes is because these are the ones that trip up the newcomer. They represent the actions meted on the newcomer that is deeply hurtful and lasting. They are the experiences that tear at the foundation of the new entrant's self-esteem. They are the ones that ambush the newcomers and knock them backward. The positive and feel-good interactions always uplift and boost the newcomer's journey and welcome their arrival and participation in the American Dream with open arms. The challenges are found in the Garys, the conferee jousting with Dr. Simon, the Pauls diminishing their bosses, and their likes. They are there. Unfortunately, they are a part of the journey and no escaping them. They are a part of the hand you are dealt as a newcomer.

Nonetheless, a few positive stories are warranted here. Gloria arrived in the United States under the visa lottery program in 2005. She settled in North Carolina and became gainfully employed as a nurse. She trained and worked as a nurse in her native country. Therefore, it was easy for her to dust off her skills and apply her trade as a nurse in America. Twenty-two months after her arrival, she felt some lump on her breast and scheduled an appointment with her doctor. Her doctor referred her to further tests, x-rays, and biopsy upon careful examination. During her regular shift at work, she received a call from her doctor with devastating news. She had an advanced stage of cancer. The diagnosis numbed her.

The doctor quickly stated that the situation was not hopeless, but they must act fast and aggressively to improve her successful intervention odds. This development crushed Gloria. She had no immediate relative within three hundred miles of her place of abode. The only family she had was her work family, which had grown larger. She shared the sad news with her boss and some coworkers. She sank into high anxiety and depression. Then her work family sprang into action. The work family's planning and execution were legendary. The group organized daily contact and interaction with different coworkers. They organized prayer meetings at her apartment twice weekly and took turns bringing prepared meals to Gloria. This practice commenced from the initial diagnosis through the surgery and unto healing and recuperation. Almost all coworkers at Gloria's work department were bona fide Americans. Gloria was a newcomer who they all rallied around with love and care at time of need. Gloria's surgery was not only successful, but she became cancer-free and has remained that way till this day, nearly fifteen years after her ordeal. Gloria later confessed that she withstood her health challenge and saga due to God's grace and the unconditional love and support she received from her work family. She continues to pay it forward.

Peter had emigrated to America as a student sponsored by his father. During his sophomore year, Peter received the sad news that his father was killed in a ghastly automobile accident in his native country. It was debilitating news for Peter, not just emotionally but financially. You see, the father was the only breadwinner in the family and funded Peter's education. As a student, Peter had to continue his education to meet his visa requirements or leave the country. He could not afford the college tuition associated with his study. The then prevailing circumstances placed Peter in a quandary. Peter shared his predicament with his local church pastor, where he had attended since his arrival in the United States. His church pastor listened intently and asked for a week to pray about it and confer with some church committee members. Then, on the Friday of that week, the pastor called Peter and requested a meeting with him. Peter rushed over to the parsonage offices, his heart pounding as if it wanted to jump out of his body. The pastor explained that he had presented his situation to the church funding committee members. Unanimously, they voted to provide his college tuition for the remaining two years of his undergraduate education. When Peter was first narrating his story and predicament, the pastor inquired about Peter's annual tuition. So, he had that figure before presenting Peter's case to the church committee. Midway through the pastor's explanation and a congratulatory note, tears began to stream down Peter's cheeks. Such demonstration of love and support was overwhelming for Peter. He had imagined that at the very most, the church might throw some stipend his way, but to conceive a full tuition ride from the church was unthinkable. They prayed, and Peter, shaking with delight and bewilderment, joyfully exited the pastor's office. Peter graduated and became a successful microbiologist at the Center for Disease Control in Atlanta.

How about this personal story? A couple that emigrated from South America lived in a neighborhood in Florida. They worked hard to purchase a modest home in a sprawling area of town. Life seemed perfect for them as they continued integrating into society. Almost simultaneously, they both lost their jobs due to the big recession of 2008. They fell behind on their bills and mortgage. Things were bleak, and nothing looked up. They occasionally held small talks with their neighbor to their left. He was a retired widower who lost his wife a few years earlier and now lived alone. He held a decent job during his working years as a manufacturing company manager in Ohio. They had moved to Florida after his retirement from his job. Whereas he would not qualify as a financially wealthy man, he was not a pauper either. On a lazy Sunday afternoon, the couple's husband picked a conversation with their neighbor to the left. Their discussion led to the financially strapped husband opening up to the retired widower about their desperate financial situation.

Without hesitation, the widower offered to assist them with a loan as they grapple with their hardship. They quickly hashed out the periodic amounts the couple needed to tie them over monthly and a future payment plan after one year, without any interest. It worked out like a gem. The couple received their monthly funds, and after one year, triggered the repayment arrangement. It became successful because five months after their agreement went into effect, the husband secured a job, and his wife followed suit with employment by an area marketing firm two months later. This display of good neighborliness was performed by the retired mid-westerner originally from Ohio.

Daily in the United States, acts of kindness and generosity occur. Individuals and organizations engage in these humane behaviors that fill big voids for assistance and acceptance. Acts of kindness are not always financial. They could be as simple as a smile, a positive compliment, holding the door open while someone's hands are full, etc. These gestures abound every day, everywhere, and almost every time in society.

"The role of culture is reflecting who we are, where we have been, and where we hope to be."
-Wendell Pierce

CHAPTER 14

REVERSING THE CHAIR

In writing this chapter, I contemplated the most effective way to illustrate the shoe being on the other foot, as the saying goes. What if native-born and raised Americans were to emigrate to other parts of the world. Such a migration could be as missionaries, business partners, spouses, or to reside. After all, many Americans frequently move to the various continents and corners of the globe. Should I enumerate and discuss the cultures and practices in these foreign lands or recount Americans' experiences that have lived in these countries and communities. I settled by highlighting the common practices and behaviors in different regions and countries and citing examples and stories of actual experiences. I reasoned that such an approach provides the reader with a wholesome understanding and appreciation of what an American newcomer to such societies might encounter and contend with. The hand that the American newcomer is dealt.

In no order, we begin with Europe. Of course, Europe represents a vast continent with numerous nations and their respective cultural nuances and bearings. These differences make Europe interesting and a rich study. In France, women do not initiate handshakes. The woman waits for the man to stretch his hand for the shake. Should the hand not be extended, the woman patiently endures until it comes. While we are on handshakes, in Poland, it is taboo to shake hands at the doorway. Oops! In Portugal, men hug or pat each other on the back in friendly settings. Women kiss both cheeks with other women and with men. The Chinese culture tends to be less openly affectionate than other cultures. For example, hugs and kisses are less common. Even shaking hands is regarded unnatural and avoided. In Chile, it is the norm to stand close to each other and considered it rude to back away. Physical contact among people of the same gender during conversation is permissible, such as placing a hand on another's shoulder. This expressive gesture signifies attentiveness and friendliness.

Back to France. In social settings, you kiss on both cheeks, and in professional environments, one keeps their blazer on at all times. It is considered rude and unprofessional to remove your blazer. In the United Kingdom, blazers are considered weekend wear and not worn at work. Americans are used to addressing people by their first name. In France, it is different. You use a person's title and last name. In Germany, you introduce yourself by your name and last name. No use of titles.

A coworker told of her first business trip to Marseilles, Southern France, by the Mediterranean. In a cavalier way, she addressed the European division CEO by his first name during a company-sponsored diner outing. The chairman, who was French, did not correct her in the midst of many but cleverly ignored her. It was not until the following day that she realized her faux pas when she complained to a fellow French coworker. She quickly sought the chairman during one of the breaks and apologized profusely. The chairman downplayed the misstep by consoling her that she was not French and, therefore, forgiven. Should you desire proper attention and focus on a business matter in France, avoid scheduling meetings or detailed reviews and deliberations during August and two weeks before or after Christmas and Easter. The lack of focus is because such periods are cherished and held dearly as vacation times.

In negotiations, various practices exist. Some cultures, such as the Slovaks and Czechs, prefer a longer negotiation process. It must have a beginning, midway, and a conclusion. Any circumvention is at your peril. In like manner, quick and seemingly rushed decisions do not win accolades in France. Thoughtful and insightful evaluations are the norm. Imagine the American newcomer who has a flowchart in mind about the speed and route of decision making. The newcomer would be frustrated and disappointed at the pace of decisions, even on critical matters. There is a general notion that the longer one ponders a subject, the better the chances of covering all possible consequences. The same could be said of Spaniards, who take their time in dealing with things. One must pack a lot of patience here. Another aspect of the Spaniards is that it is not considered rude to interrupt when speaking or be interrupted. It would be acceptable and permissible to interject in the middle of a discourse.

Let us examine the hand and fingers' unique symbolism and signs. In Germany, thumbs up is the sign for "one." Africans never use their left hand when giving gifts. They use either the right hand or both hands. Receiving an item with a left hand signifies disrespect, especially if the giver is an older individual. Also, in handshaking, a common practice involves starting with the person to your right to the left. As a sign of respect, Africans bow their heads to superiors or older people. Within some parts of West Africa, a stretched-out open palm directed at you represents cursing almost equivalent to the American version of "giving the finger." In the Middle East, holding fingers in a pear-shaped configuration with tips pointing up at about waist level and moving the hand slightly up and down means slow down or be careful. This gesture is easily seen on the cities' crowded streets, usually emanating from impatient taxi drivers or frightened pedestrians. It is considered rude to put hands in your mouth in China, biting your nails, and removing food from your teeth. Additionally, the Chinese do not pat others on the head. The head is considered a most sacred part of the body and, therefore, venerated.

Food is an essential part of life. So, let us visit some food and cuisine practices in some countries and communities. In Africa, you do not begin eating unless the eldest man has started eating. This practice is in deference to age and, supposedly, wisdom. Different cultures regard certain food items as delicacies. For example, In China, scorpions, locusts, snakeskin, dog meat, and blood may be presented among other menu items. In France, foie gras (enlarged duck liver) is a common and enjoyed food item. In Thailand, insects served different ways are consumed as street food. Leaving unfinished food on your plate is considered rude in Chile. By not devouring every bit on your plate is interpreted as dislike of your meal. Besides, not staying after meals to engage in small talk and conversation is equally regarded as disrespectful.

Conversely, in China, eating everything on your plate is termed that you are still hungry and deemed offensive to your host. In the middle east, traders and street food and souvenir hawkers tend to be persistent and get in your face and space. They hustle, trying to make a living to take care of their families. The family is seen as a person's ultimate refuge and support system and defended fiercely through honor (self-respect to self-pride). Shame to the family is avoided at all costs. Besides, public display of affection between two individuals is frowned at. People tend to be more hospitable than other cultures. This habit has been traced to welcoming nomadic travelers who came in from the harsh desert conditions and were treated with food, water, and shelter. Many hotels conspicuously practice enhanced hospitality to this day.

Let us review numbers. In China, number 4 is akin to 13 in the west. Fourth-floor rooms are avoided in the same way as the thirteenth floor in the west. One must avoid scheduling meetings at 4 pm or on the 4th day of the month, especially in April. Number 4 is superstitiously a symbol of bad luck. Four is a taboo number because it sounds like the word "death" and is considered unlucky. Also, meeting seating is methodical and hierarchical in China, with the leader sandwiched among the attendees. Also, meeting rooms are entered in the same pecking order.

In Italy, maintaining eye contact is the norm, and flirting, especially by women, is acceptable and permissible in business or social settings. On his first international assignment to Milan in 1998, Tom thought he had landed an admiring dame when a company liaison, who was Italian, flirted with him. Tom was clueless that the overtures extended to him by his Italian overseer were simply courteousness to make him feel welcomed and at ease. He interpreted her pleasantness as perhaps a romantic inclination. With such a misguided mindset, Tom made his move by inviting her for a private dinner. The rejection was equally flirtatious. She smilingly told Tom that she could not oblige due to some commitments. Tom became even more confused because she did not vociferously turn him down. In Tom's mind, she still harbored some likeness but letting him in gradually. What was more, she would inquire of Tom's previous night's comfort and sleep every business day morning. She really cared and just wanted not to rush matters; Tom rationalized. As Tom persisted in his delusional escapade, the Italian hostess broke it down for him. She invited Tom for coffee at a nearby café and explained that, unfortunately, he had misread her goodwill gestures towards him. The amiability was to help him adjust to his new environment, nothing more, nothing less. It dawned on Tom that he was in a new land, where behaviors, utterances, and attitudes were different from his native America.

A Christian missionary family told their experience in migrating to a new community with their different cultures and practices. The husband and wife had been assigned to a post in West Africa. They traveled with their two children, Carter, nine years, and Kimberly, seven. After an exhaustive twelve-hour flight, they arrived at their destination and were picked up by the local parish's volunteer chauffeur, who drove them to their living quarters. Their planned residence had been thoroughly cleaned, adorned with fresh bed linens, curtains, furniture, etc. It was spotless and airy. The parishioners were excited about their sojourn and religious work among them.

They arrived at the mission campus and unpacked their luggage from the vehicle. The local church leader welcomed and led them into their residence. He showed them the cottage, comprising a small living room and two bedrooms with a bed in each. When they got to the supposed children's room, Kimberly innocently observed loudly, "There is only one small bed here. If Carter takes it, where am I going to be sleeping"? The local church leader attempted to explain that she and the brother, Carter, share the same bed. At this juncture, Kimberly stormed out of the room in tears. To fully understand the dynamics of what transpired, one needs to delve into the host community's culture and tradition. In West Africa, the family unit is very cohesive and tight. Many children would unassumingly sleep on the same bed. A full-size bed or mattress would accommodate three or four young individuals. This practice is commonplace. Only occasionally are children assigned their rooms or beds. Parents institute this mindset, and children's expectations are calibrated accordingly.

On the other hand, Kimberly was conditioned by her native country's norms to possess her room and bed. No fault of Kimberly. Different societies, different practices. Are you curious to learn how this "major catastrophe" (Kimberly's term then, according to her Mother) ended? Children and young adults adapt to change better than older individuals, social scientists tell us. Kimberly's experience with this matter added one more validation to the scientists' assertion. Suffice to say that Kimberly quickly turned a different page once her parents thoroughly explained the situation, and she made new friends at school and her new community.

A plastic molding company in Florida established a partnership in Shanghai, China in 2001. As a part of the agreement, a co-manager was relocated to the plant from the United States. Fred, the relocated manager, had briefly visited China on a family vacation. He and the entire family moved to Shanghai. Before their trip, Fred had bought a couple of clocks to be presented as gifts to his new coworkers. Thus, began Fred's rapid-fire acculturation, or should I say misfire into the Chinese culture. During his first week in China, Fred made two missteps in the Chinese culture. The first was the clock. Clocks are one of the items tied to death and separation in Chinese folklore and customs. The others are shoes, which sound like evil or heretic and signify wanting the gift receiver to walk away from you, thereby ending a relationship. Never give pears because it sounds the same as separation. White flowers are also a no-no because white and black represent mourning colors.

Fred wrapped the clock gifts and excitedly gave them to two Chinese co-managers. They would not open their gifts in front of the giver, per their custom. So, they took the gifts home. The following day, both gift recipients uncoordinatedly returned them to Fred. They explained they could not accept the gifts because of their symbolism and thanked Fred for his kindness and thoughtfulness. Lesson learned!

The second learning occurred when Fred had taken his direct reporting employees to a social dinner outing. One of these employees had recommended the restaurant, which was a popular local watering hole. The meal was delicious and the service impeccable. When the waiter served the bill, Fred proudly tendered the total plus a sizable tip. All in cash because credit card utilization, especially in non-tourist facilities, is minimal or non-existent. The waiter collected the amount and left. After about five minutes, she returned and handed the tip amount back to Fred, insisting that he had overcounted the local currency. Fred attempted to explain that the extra amount was tip-money for her excellent services and stewardship. Through Fred's dining mates, she quickly countered that tipping was not the norm and practice in their community. She jokingly added, "Tipping is American, and we keep it in America." These two experiences within Fred's first week in Shanghai demonstrated the imperative in learning and adapting to your new community's way of life.

In Norway, dining represents a nearly sacred activity. For instance, at restaurants and other eateries, diners are never rushed to turn the tables. You are allowed to eat at your pace, and,if needed, hold long conversations thereafter. Waiters provide the space and latitude for complete relaxation. The bill is not rushed to you. In practice, the bill is not brought until requested by you, regardless of the time lag. The servers are trained not to signal any pressure in pushing you out. The culture believes that food must be thoroughly enjoyed and venerated.

Retired English professor Alan and his also retired sociologist professor wife, Margaret, decided to live their retirement lives in Chile, South America. They had vacationed there a decade ago and fell in love with the land, its people, and offerings. A part of their draw to Chile was the country's affinity for literature. Chileans call their nation the "country of poets," and Chilean poet Pablo Neruda received the literature's Nobel Prize. So, Alan and Margaret were all in. They packed and moved to Valparaiso, incidentally, poet Neruda's former city of residence. Margaret was a beautiful lady even at an older age. She was in her early sixties but could pass for a forty-something attractive female. Besides, Margaret was a health buff and took her nutrition and exercise regimen very seriously.

Numerous times on outings such as dinner, men stared at Margaret, much to her embarrassment and confusion. A few times, the stares were intense that she would ask Alan if anything was amiss with her clothes. One day, a Chilean botanist whom they had befriended visited them at their hillside home. During a casual conversation, Alan complained about the stares at Margaret by men. Their friend busted out laughing. He quickly explained that a Chilean woman would die for such stares. He continued, "In Chile, it is considered positive flattery for a woman to be stared at by men. It is admiration and validation of beauty." He turned to Margaret, who was shaking her head in apparent disgust and said, "Our men are saying, you are beautiful!" Alan and Margaret never could wrap their heads around that practice. However, it represents the custom and culture of your new land. They became accustomed to the stares that sometimes Margaret would wave at the starer, who would reciprocate with a smile. Another adjustment Alan and Margaret made was to discontinue referring to the United States as America while living in Chile. Should one insist on inserting America, remember to add North. Chileans regard all of South America also as America and frown at the usurping of America's designation by the United States.

Traffic jams and associated stressors in different countries and cultures deserve mention here. Motoring captures the culture, norms, and practices of the society in its unique way. Observing how citizens conduct themselves in this activity demonstrates their values and guardrails. If you have visited major world cities with teeming populations, you will recognize the traffic issues and headaches inherent in driving in them. Usually, a rising economic lot of citizens results in increased car ownership. With more vehicles on the road and greater urban development, existing transportation infrastructure becomes overburdened. These factors lead to congestion, pollution, and road safety challenges. In some of these cities, you drive at your peril. There are no rules. Pedestrians' road crossings are at their own risk as they dash and dodge the impatient drivers and oncoming traffic. Sometimes, traffic snarls can last up to four hours at a time. Unbelievable! A newcomer to this environment must pack a boatload of patience and perseverance.

Let us discuss far away funerals. Clara was born and raised in Kansas. She is as Midwestern America as there is. While in college, she met and eventually married a schoolmate named Phillip. Phillip was born and raised in Africa and came from an upper middle class family. The father was a medical doctor by profession and the traditional village chief. As the chief, which could be equated to a town manager in an American sense, he enjoyed the respect and wielded a lot of power. However, he never abused his power and was beloved by his constituents. He took ill and was felled by the sickness. Phillip and Clara reside and work in Portland, Oregon. Arrangements for Phillip's father's burial and the funeral had to be made. Clara's initial culture shock was that Phillip's father's body was to be preserved in the mortuary for about six months. This delay in the burial and funeral allowed for adequate planning and the return of all of the deceased children scattered worldwide. Two of his children live in America, one in the United Kingdom and another in Australia. Though the reasons for the delay seemed cogent, Clara could not quite understand keeping a corpse for that long before burial.

Then it was time for the burial and associated events. Phillip and Clara flew to Africa for the activities. As they got there, Clara learned that the obsequies were slated for four straight days. Normally, it would consume two days but as the village chief and a beloved one, the village wanted to demonstrate their love and affection for their deceased chief. The length of the events was another shock for Clara. As Clara would tell the story, these earlier twin shocks were minor compared to what was to come. On the first day, church services were held at the local sanctuary with more than twelve priests in attendance. The bishop of the diocese was present. The service lasted forever, according to Clara (it actually was four hours). After service, the crowd, numbering close to five hundred, headed for the town hall where the chief's remains lay in state for subjects to pay their respect. That was just day one.

On day two, the chief's remains were brought to his residence. Here, a smaller religious service was held, and the chief was duly buried. After the burial, various groups with their regalia would either march or dance into the chief's large compound to honor the late chief. Over five different troupes came on that second day. Then came the third day with more groups. Each group was trying to outdo each other with their attires' colorfulness and their celebrations' boisterousness. Clara was wonderstruck. She maintained that the only event she could compare it to was the Mardi Gras in New Orleans, a mini-Mardi Gras, of course. Still more groups came on the fourth day, which marked the conclusion of the funeral. As discussed in the earlier chapters of this book, death and funerals are hinged on culture. Phillip's father's example represented a celebration of his life on earth. Though they grieved his death, they emphasized his living well and respectably while with them. Since we all will die, they send the dead off to the beyond jubilantly. Those with a different cultural perspective could misinterpret the celebration as a good riddance for the deceased when it represents bidding goodbye to the deceased with a bang.

In an undergraduate class decades ago, we learned, albeit briefly, about the Eskimo (comprising the Inuits, Aleuts, and Yupik tribes) people and their customs and traditions, including wife-swapping. To the Eskimo, offering one's wife to a guest represents extreme hospitality cloaked in spiritual and practical conditions. However, researchers have debunked that Eskimos offer their wives to strangers. Instead, it is validated that Eskimos swap wives among themselves after a séance (spiritualist meeting to receive spirit communications) performed by a Shaman (practitioner dedicated to healing, divining, and forecasting with connection to a spirit world). These activities are ritualistic rather than sexual deviancy, as most westerners view them. The western mindset negates cultural relativism and judges the Eskimo practice from our cultural lens.

While we are on people and geographical locations, let us discuss a few more. In Northern Spain, men dressed as the devil run between and jump over the newly born within the year laid on a mat. This tradition is commonly known as "baby jumping," and a way to clense a new baby's soul. This annual practice has occurred since the 17th century. The Danes ring in the new year by throwing dishware, usually broken, at the homes of friends and family to wish them good luck for the year. I imagine the recipients have a lot of fun cleaning up after the piled broken dishes, plates, cups, and bowls. A little sarcasm does the soul good!

How about this one? The Toraja people of the Sulawesi region, Indonesia, live with the dead bodies of their relatives for some months before properly burying them. I will not ask about the possible stench because such a culture usually possesses a way to mitigate any possible odor emanating from the bodies. Some cultures in Africa bury the dead inside and below the bedroom floor of the next of kin. Yes, the room is torn apart and the grave is dug inside and redecorated after the burial. The next of kin supposedly lives with the body and soul of their loved one. Ponder this one. The Romans believe in feeding the dead. Therefore, they install pipes in graves at the burial grounds through which the kin of the dead pours honey, wine, and other food items into the graves. These examples demonstrate the relativism of culture and norms.

Whoever thought that sunbathing and tanning could present a cultural dilemma? Well, it did in far away Guyana. A coworker born and raised in this tropical country told of his experience with sunbathing and tanning westerners. As teenagers, he and three others were hiking a sparsely inhabited area. Suddenly, they saw an almost naked couple in a makeshift park. With very skimpy underwear, they swore the couple was naked from their vantage point. They retreated and went straight to one of the parents to report two seemingly insane individuals without clothes confused and laying out in the brush. The alarm they raised caused the parents to investigate the information immediately. As the parents approached, the couple reached for their towels and covered themselves. Upon further inquiry, the couple were Australians who had recently relocated to Guyana. They attempted to explain they were engaging in the conventional treatment of their Caucasian skins. Pretending to understand, the parents walked away but in murmured disgust how one would purposely indulge in roasting their God-given skin. Suffice to say that sunbathing is not for Guyanese.

These Guyanese did not know and appreciate that tanned skin is an aspect of beauty in some cultures. In western cultures, tanned skin signifies upper social status. A tanned individual is deemed to possess more free time and disposable income to spend outdoors rather than confined to the home, office, or place of work. Therefore, the sun-kissed beach look is appealing and adored. People proudly expose their tanned skins, especially after vacations. Invariably, tanned skin portrays time at some beach or exotic locations. However, the treatment and regard for tanned skin were different in western societies a few centuries ago. At those times, people that spent more time outdoors were typically field hands. Lighter skin was more desirable since it depicted a higher social class that comfortably allowed one to stay indoors.

Hopefully, this chapter shows that as Americans, we are susceptible to the stark realities of migrating to a different country or society with its varying norms and customs. Perhaps, it magnifies acculturation and assimilation to new communities, as the chair reverses.

In conclusion, please indulge me in pondering the impact of the COVID-19 pandemic on some of these norms and practices. The pandemic affected every nation in the world, though, at varying degrees. With its spread, people and communities adjusted their lifestyles to slow its propagation. Actions taken include, but are not limited to, lockdowns, social distancing, mask-wearing, remote working, and allowing only small gatherings of people. While it currently seems that the pandemic's spread is slowing, anxiety levels are still high. So much so that people are reluctant to engage in some pre-pandemic norms and practices.

For example, with handshaking, cultures that emphasized this ritual may need to reconsider its practice. Cultures that cherish embrace, hugs, and kisses on the cheek may rethink and perhaps modify these norms. Online meetings, learning, and remote working will enjoy a heavy boost post-pandemic. Even buildings and structures, such as schools, auditoriums, and sports centers may experience new designs and layouts. Building architects will design structures to minimize the spread of any virus through the air vents. It will be interesting to observe how various industries and occupations adjust to future fear and concern of viruses and communicable diseases. How much elbow room would Americans need for personal space? Would such distance be forever elongated compared to the pre-pandemic standards? Or would it return to its prior levels? COVID-19 undoubtedly has shaken our collective psyche and could rearrange some of our attitudes, behaviors, and way of life, albeit marginally.

The most impactful area post-pandemic will likely be in remote working. Remote working is a trend that continues to gain personal and organizational converts. The COVID-19 pandemic accelerated the embrace and deployment of this feature within the workplace. Global Workplace Analytics estimates that by the end of 2021, 25-30 percent of employees will be working remotely, at least, in part, compared to the pre-pandemic level of about 16 percent. Moreover, most newer generation members prefer jobs they could perform either wholly or partly remotely. What is more, organizations are discovering the direct and indirect benefits of remote working. The pandemic imposed widespread remote working on society. However, it has become a more significant part of the workplace ecosystem and will gain further ground post-pandemic.

"No tattoo is made without blood."
-Mozambican Proverb

CHAPTER 15

SO, WHAT NOW?

Do you dig a hole and hide, or maybe crawl back into your cocoon and wallow in self-pity and rejection? Of course, not. You must shed the drawbacks and hurdles that these negative encounters ferment in your psyche. You must place these experiences within the backdrop of one's total exposure to a new land. Because, overall, the positive and enlightening jewels far outweigh the few setbacks and nuisances. You latch on to the fundamental American goodness that you have received. You bring those memories and thoughts front and center in your mind and heart. It is such rekindling that sustains you during the rough patches in your assimilation into your new land. At the same time, you garner some learning and grit from these realities.

Though not scientifically documented, it has been anecdotally observed that one's degree or level of immersion in your new land's culture correlates with one's level of assimilation. So, suppose you recoil due to some emotional or psychological knocks prompted by the receipt of hurtful or demeaning treatment. In that case, you are likely to prolong your embrace and needed understanding and appreciation of your new land's values, culture, and way of life. You are, therefore, encouraged to plunge head-on with reckless abandon into your new environment.

The rest of this chapter will be devoted to stories, instances, and statements highlighting approaches and methods of navigating some of these landmines along your acculturation journey. These do not represent an exhaustive list or methodology. Instead, depict some examples to whet one's appetite and curiosity and perhaps provide some key takeaways for the reader.

Justice could not be done to the topic of assimilation without, at least, a cursory examination of what assimilation is and entails. Sociologist Milton Gordon, some decades ago, submitted that assimilation is multidimensional and complex. He postulated that assimilation has many components encompassing economic, social, cultural, and political angles. Sometimes, these dimensions are not in lockstep and synchronized. They can be disjointed and chaotic in practice. What may make political sense for the newcomer may be at odds with their cultural, social, and economic conditions.

Assimilation is fraught with internal conflict and tension. Scholars have argued extensively about what fuels accelerated assimilation of an ethnic group. One school of thought holds that the isolation, whether by choice or circumstance of these groups from the mainstream community, causes delayed or even incomplete assimilation. Conversely, others have advanced that lack of assimilation is not isolation, rather the weakening of boundaries and barriers between groups. They submit that opportunities for more interaction also result in more avenues for conflict. We will not attempt to settle this argument here.

Assimilation is not a one-way proposition. It could represent a forward process that is reversible. For example, fully assimilated citizens who now speak fluent English might choose to expose their American-born children to the old country's language. This reversal often stems from the fear and concern of losing one's heritage. Some native-born believe assimilation cannot be complete without the total abandonment of all traces of one's heritage or vestiges of the old country. Partial surrender is viewed as superficial and disloyalty to the new country that nourishes you now. After all, if you do not stand tall for her, why migrate here in the first place, they argue. Little wonder some native-born citizens sometimes view hyphenated American groupings such as Italian-American, Irish American, or African American as an affront to Americanism. If there were ever a rub, it is in the native-born citizen's uncompromising belief and assertion that all newcomers must be pure in their subjugation to all things America. At the same time, the newcomers desperately attempt to cling to some sliver of their heritage tenaciously. As one sheds some aspects of the old country's culture and embrace and inculcate certain values and culture of the new land, the individual's internal tension and conflict elevate. This predicament becomes a period of extreme vulnerability for the newcomers. This equivocation is because they are just culturally knocking at the American door and simultaneously uprooting the foundations that made them who they are. This phase usually is the most agonizing and heart-wrenching period for the newcomer. The experience of this period is also individualistic in terms of degree, speed, and duration. For some, it is a mere few years. For others, it could take decades, and for the rest, assimilation may never wholly occur.

Assimilation in America is comparatively more flexible and accommodating than other cultures and nations. This flexibility and elbow room have been touted as the main reason for the more successful newcomers' acculturation in the United States. The expectations of assimilation in the U.S. do not require that newcomers completely repudiate, overtly or openly the old land's cultures. This assimilation approach led to the naming metaphor of America as the "melting pot." The melting pot concept captures the essence of diverse ethnic groups' intermingling with varying cultures into one national identity – America. Some have argued that the melting pot metaphor had lost its luster if there were ever one. They submit that the terminology is incongruent with the realities of ethnicity in the American ethos. If melting were the operative word, some ethnic groups' peculiar cultures never totally melt and become indistinguishable. Some remnants persist, and some argue, should. Maybe that is why sometimes America is described as a mosaic of ethnic groupings or a kaleidoscope. Writer and social scientist Ronald Fernandez recently coined the metaphor "banquet of cultures" to describe America. Whatever position one takes on cultural pluralism, one thing is certain: to exist and properly function as a nation with a relatable identity, each ethnic group must eschew some aspects of its original cultural underpinnings to create the semblance of a national persona.

Social Scientists utilize four fundamental yardsticks to measure and assess assimilation into the dominant American culture. These are: socioeconomic status defined by educational attainment, occupation, and income; Geographic or spatial distribution and concentration, which involves residential and geographic patterns. Language attainment addresses the ability to speak and be fluent in the English language. Intermarriage involves marriage across racial, ethnic, and generational lines.

Isabella spoke some English before migrating to the United States from South America. She could hold a very light conversation in English. All of her secondary and post-secondary education was mostly in Spanish. Now Isabella is living in the heart of Los Angeles with some language barrier but full of energy and zeal. She was resourceful and through her church, accessed a non-profit agency that teaches English as a second language for free. This program was operated and populated by retired area volunteer former English teachers who were paying it forward. Isabella enrolled in the program, and within eleven months, she was speaking English fluently and writing semi-professionally. Similar programs abound nationwide.

Children generally adapt to their new land much faster than their parents and guardians. Let your children loose to plunge, though, with expected parental guidance. Then "follow" them. What following means is to intentionally expose them and yourself to the community, its facilities, and its institutions. Engage in sports and neighborhood events and activities with them, and possibly volunteer in some of these endeavors. Nadia, her husband, and two children emigrated from Eastern Europe and settled in New Hampshire. Her husband was gainfully employed as an engineer at a local manufacturing plant, while Nadia was a stay-at-home mom for the first two years of their migration to the United States. She had a lively and vivacious personality and always bustling with energy and high spirits. But these were all bottled up in their modest three-bedroom apartment home. At the local church they attended, Nadia conveyed to one of the older women how bored she became after her husband went to work and the kids were off to school. The elder lady suggested she volunteer at a church-run senior center not affiliated with their church. After some exploration and mutual evaluations, the senior center accepted Nadia as a volunteer, working four hours a day and three days a week. This deployment provided the outlet Nadia needed to blossom. She was a "hit" at the senior center from day one. Her effervescent personality, combined with her affability, made her an excellent addition to the staff. It became so mutually beneficial that Nadia extended her days to four days per week.

This engagement was the tonic Nadia was yearning for in her assimilation journey. Her daily interactions with staff, residents, and visitors accelerated her English language proficiency. Her work exposed her to more American values and culture and taught her some subtle American ways of life. She confessed that her volunteer work at the center did more to her than even an organized school.

Hans was promoted and transferred to his employer's U.S. location in Minneapolis from the Netherlands. Hans had occasionally vacationed in the U.S. with his wife and their young son, so they liked the U.S. and its offerings and had lobbied heavily for the open position and were equally qualified to assume the responsibilities. Hans and his wife knew what they were bargaining for by emigrating to the United States, or so they thought. As they settled in their new land and home, they began to discover that the treatment meted to them as tourists and visitors was different from soon-to-be fellow citizens. It was no longer "welcome, enjoy yourselves, and goodbye." They are here to stay. Even among coworkers, Hans observed that the reception and interactions became somewhat tempered and guarded. They are viewed as direct partakers of the American pie and dream.

Hans and his wife were determined to live out their American dream and plotted they would not be denied the opportunity. Their chosen path was to "bulldoze" their way into things by belonging to as many social, civic, and community clubs and associations as they could handle. That paid off handsomely because they cultivated a vast and expansive friendship network that formed a strong support system for them. They now brag that you can essentially control your destiny in terms of assimilation into the dominant culture and way of life in America and tease that they could make money advising newcomers how to acculturate appropriately and successfully.

Cultivating or maintaining a strong work ethic represents one of the pillars of American values and cultural foundations. It is through hard work that one pursues and attains the American dream. So, to have a fighting chance for your dream, one must subscribe and live by the credo of a strong work ethic. Undoubtedly, this proclivity aids the individualistic bent of the American culture. You can become whatever you wish to be as long as you roll your sleeves and go to work.

Jekwu hailed from a comparatively wealthy African family. The grandfather had started the wealth legacy by being a local land speculator, real estate investor, and successful bakery owner. Jekwu's father took his father's solid economic foundation to a higher level by expanding the family's portfolio of companies and investments to include oil and hospitality entities. Locally, the family was regarded as perhaps the richest dynasty. Jekwu was the eldest son. He was sent to further his education in the United States. Jekwu's intention all the while was to emigrate and settle in America. He was exposed to America's trappings, including an organized society with superior consumerism through the yearly summer family vacations in various parts of the United States. The American society and values resonated with Jekwu immensely since he had enjoyed the proverbial silver spoon in his mouth all of his life. The father would not hesitate to lavish their wealth on Jekwu, believing he needed to be adequately catered for as the heir apparent. Jekwu became spoiled due to this pampering and lost his way regarding a strong work ethic.

For the first couple of years that Jekwu emigrated to the United States, the family money was flowing non-stop to support him. He had never held a job in his life. Yes, occasionally, while in his old country, he would make stops at their numerous company locations to say hello to some relatives and family friends dutifully at work. So, initially, while in the United States, he did not consider taking up any job, even during the long summer breaks from schooling. In his second year in America, Jekwu's father died. The younger stepbrother (Jekwu's father had three wives), their late father's understudy, took over the family business rein. However, there was a slight problem. Jekwu and his stepbrother never got along. They were like water and vinegar. All along, the unnecessary waste of family funds on Jekwu overseas was a major source of irritation to him, the stepbrother would complain. So, when he rose to the head of the business empire by the father's death, he immediately choked off the monthly funds to Jekwu. Let us put some clarity on this. The stepbrother did not curtail the funds in their entirety, instead slashed them considerably, arguing that whatever was being remitted monthly ought to sustain a middle-class and comfortable lifestyle. The problem was that Jekwu had become accustomed to an ostentatious living that he could not tone down his tastes and preferences easily. After all, he did not know any other way all his life. He never cultivated a strong work ethic, and he was now in a society that encourages, cherishes, and revers this virtue.

Jekwu had to adjust and pivot quickly. After numerous phone and written squabbles with his stepbrother, which did not yield any additional funds and introspections, Jekwu concluded he must start fending for himself through hard work and self-reliance. Jekwu had now arrived at the American values and culture shore, belatedly, of course. Though, through a circuitous path, he got there. Anchored in this new reality, after about three years, Jekwu became a consummate American citizen rife with all the virtues of Americanism. So much so that Jekwu became a strong proponent of self-reliance and hard work. Ironically, he maintained and gave his stepbrother a back-handed thanks and compliment for "forcing" him to wake up to self-reliance and independence, both in thought and actions. This story had a happy ending, which I must share. As Jekwu visited with his family in Africa occasionally, the stepbrother observed and complimented him on what he termed an elevated and sophisticated worldview, which among other things, espoused self-reliance. The stepbrother confessed that what had driven his earlier resentment and anger was what he regarded as a lack of self-reliance. He saw Jekwu relying on and milking the family wealth recklessly, he reasoned. With Jekwu's attitudinal shift on this matter, the two brothers reconciled and became best of friends, in addition to being brothers. They are much older now with grandchildren, and their relationship remains strong as ever. When they tell this story now, they chuckle and say that the youthful hormones were errant with both of them during those tumultuous years. Their children can tell this story as much, if not better, than the two patriarchs. This learning is because they both decided to use this episode and saga as a teaching subject to the children on family unity and self-reliance. The children heard them loud and clear because they are all go-getters who chart their courses in their different professions.

Any realistic newcomer would not initially expect parity with the native-born regarding access to jobs, salary, and wages. Over time, which in some cases may cover a generation, equivalence is achieved. The new entrant's opportunities are dimmed by certain drawbacks such as language barriers or accents, reliable transportation modes, unfamiliarity with the cultural and social landscape, and subtle and sometimes overt discrimination. Newcomers accept jobs that are below their qualifications to establish and gain a job beachhead. In the late nineties, I had traveled to Washington D.C. for a business meeting. As I landed at the Reagan National Airport, I hired a cab to take me to my meeting location. The taxi driver was well-groomed and eloquent in his initial greetings and welcome. After driving for about seven minutes, he politely struck a conversation by inquiring about the city I flew from. I responded, and we continued our discussion most of the trip. The driver had revealed that he and his family recently emigrated to the country on the Diversity Lottery Program. He was a college professor with a terminal degree in his old country. To support his family and avoid dependence on public assistance, he chose to take on any menial job at this period in their emigration and assimilation journey. What was striking to me was his pride and avowed understanding of his decision and choice. He fully understood that he could not successfully compete for a professorship position at that moment. That pursuit would be for a later day. He explained that he needed to quickly come out of the chute establishing the rudimentary and preliminary foundations of life in a new land, namely, history of responsibility, self-reliance and hard work, credit building, and accountability. In my opinion, then and now, he got it! He seemed on his way towards successfully attaining his American Dream. This belief is because he was taping into the American way's basic tenets by his actions and attitude. Barring any trip up, he had a high probability of succeeding and thriving in his new land, I reasoned. My attempts to track him down through the cab company he worked for failed. I would not be surprised if he is currently teaching at some institution somewhere in the United States and enjoying his new land's largesse and opportunity.

Or, how about the waitress in a Los Angeles fine dining restaurant, who had emigrated from Slovakia. Adriena, popularly addressed as Adria, was a trained occupational therapist in her native Slovakia. She had worked in that field for over eight years before she emigrated to America. Upon her arrival, she unsuccessfully sought jobs in her field, only to find that she was required to undergo additional or complementary education before she could deploy in her professional sphere of occupational therapy. Funds for education were not readily available and accessible, hence accepting a waitress job to make ends meet. Prevailing circumstances condition most newcomers to adjust their expectations to the realities in their new land.

As an incidental topic to self-reliance and hard work is the subject of public aid and assistance. Public support and assistance are sometimes required to lift one from a particular situation or condition that may be fleeting or last longer. However, as a newcomer, you should strive to wean yourself away from public assistance, if possible. A perpetual reliance on public aid that can be overcome through hard work subjects one to a conditioned state of dependence. Observe that the operative term here is "public assistance reliance that one can overcome with hard work and self-reliance."

In chapter 2, we discussed the consumerism and credit reliance of the American economic system. This connection is real. Society is built on this premise and fundamentals. You cannot avoid it. You cannot shoo it away. You simply cannot escape it. Therefore, you must find a way to embrace and leverage it. One of the quickest ways is to start establishing and building your creditworthiness. You start slow and network with friends, relations, and institutions to guide you through the credit building process. This process allows you to pursue and eventually attain your American dream. You will be tempted to cling to your "only buy what you have cash for" mentality. Realize that whereas you could stick with such a mindset, it can be very limiting and hampering as you start to yearn for bigger prizes in your quest for the dream you are chasing.

Allow me to discuss an item that sometimes is misunderstood or misconstrued. You may erroneously believe becoming a true American means cutting off your cherished personal connections to your old country, including friends, relatives, and acquaintances. This idea is far from the truth. It is not a contradictory practice or stance to your desire to become a full-fledged American. Intriguingly, the notion and practice of staying connected to your ties in your old country make one appreciate the beauty and virtues of one's new land. This valuing is so because, with such connections, you are aware of and knowledgeable about the goings-on, including negative events and stories that buttress your decision to emigrate. These stories and activities reinforce the quest that set you out on this life's journey and adventure.

Socialize and watch the news and programs on TV. It is generally believed that TV and other media present the fastest avenue for integration into a community's culture. One thing is certain, TV and programs depict and reflect the societal culture. Binge TV watching is not the suggestion here, rather reasonable and moderate TV watching to understand and appreciate your community.

Mariam emigrated from the Middle East. She holds a job as a Certified Nurse Assistant at one of the nursing homes in New Jersey. She emigrated to the United States at age thirty-two. This migration age is important because Mariam was very entrenched in her native land's culture before coming to America. It would not be easy for her to shed the culture and values of her old country quickly. This challenge was more pronounced in some of her attitudes, beliefs, and biases. For over three years living in America, Mariam rarely watched American news or themed shows and programs. She subscribed to the cable channels of her old country. She knew and absorbed more of what was happening in her old country than where she lived. This behavior was not only troubling but counter-productive towards her integration into the American system. No one would deny Mariam of the connectivity with her old country. However, Mariam could not successfully assimilate into her new land by consciously tuning out her new abode's activities and tenor of happenings. She was now in the United States, and she might as well plunge rather than standing at the river's edge.

To illustrate this deficit, Mariam was always at a loss in understanding and interpreting American colloquial language and slang. Mariam's coworker hysterically told of her intense struggle in understanding and appreciating some American sayings. She relayed how Mariam rushed to her one day in confusion that another coworker, upon her telling her that her daughter was competing in a spelling bee contest that weekend, told Mariam to tell her daughter to "break a leg."

Mariam not only thought that it was rude of the coworker to suggest such but criminal. "How can she suggest that my daughter break someone's leg?" Of course, the coworker merely wished her daughter a great performance. Stay with me as I retell another one from the coworker. A friend offered a ride to Mariam one day and had casually told Mariam that she would have to ride "shotgun" since she had other stuff in the back seats. Mariam hyperventilated, arguing she would not have anything to do with anything gun or gun-related. Mariam had emigrated from a very violent area in her old country where guns were rampant. Having grasped Mariam's anxiety and misunderstanding of the slang, the would-be driver broke it down for her by explaining that it meant she was to ride in the vehicle's front passenger seat.

Please allow me to share one more of Mariam's enigmatic dalliances with American lingo. Mariam had called into her place of work before starting time to advise that she was sick and therefore would not report to work that day. A company receptionist took the call meant for Marian's supervisor. A call that would ordinarily last for a minute or so took an entire four minutes. That was because as Mariam spoke, and with her accent, the receptionist could not fully understand what she was attempting to convey. As the receptionist inferred that she might be calling in sick, she inquired, "Are you under the weather?" Mariam was clueless about what the receptionist asked and kept repeating, "What? What? What?" It quickly dawned on the receptionist that Mariam did not understand her question and broke it down further by asking, "Are you sick? It was then that Mariam shouted, "Yes, very, very, sick."

The moral of Mariam's saga is simple. As a newcomer, you isolate yourself at your peril. You can graduate your acculturation at your preferred pace and comfort but never strive to exist in your bubble. Avoid secluding yourself to the detriment of your assimilation into your new land.

In addition, expose yourself to the surrounding technologies and digital apparatus and devices available or that you can afford. Enhance your skills and competencies in the technological and digital world around you. Many charitable and non-profitable organizations, including some churches, provide free basic computer training and courses. Finally, tap into agencies and organizations that offer guidance and resources to new arrivals. Some of these organizations maintain robust and helpful programs and curricula.

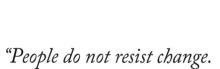

"People do not resist change.
They resist being changed."
-Peter Senge

CHAPTER 16

OH NO, WHO AM I NOW? HYBRID SELF

Any newcomer to a new land undergoes some form of metamorphosis. The clash of seemingly two cultures results in some blending and fusing of all exposed cultures. You are no longer who you were before you emigrated to the United States, but you are not totally and absolutely the embodiment of your new land. This split personality of cultures occurs because one does not and could not wholly shed all ingrained traditions, practices, and mindset. Despite the immersion in the culture of your new land, individual attitudes persist. However, as we are reminded in the Bible (Ephesians 4:22-24), we put off our undeserving and incompatible ways, perspectives, and grounding. This tug of cultures and values eventually gets resolved by adopting a balanced state. In some cases, the cultural tensions are never entirely resolved.

Some newcomers have attempted to select and live by what they perceive as the best of both cultures. Historically and practically, this approach muddles the water and creates more challenges than it solves. This unsettledness is because culture and cultural underpinnings have contexts. Cherry picking specific cultural components and applying them without the relative backgrounds render such applications superficial, meaningless, and confusing.

A friend, who emigrated to the United States about thirty years ago and retired from teaching at a university, proffered an assimilation approach for newcomers. He believes the new entrant should inventory both cultures and quickly embrace the dominant matches while gradually adjusting to the outliers. He explained that you become objective by methodically and sensibly sorting these major cultural foundations and avoid sentimental hang-ups. He submitted, though, that no matter how thorough and organized one handles the cultural clash, it can never be devoid of tension and confusion. The big question becomes how you absorb and incorporate your new land's values and culture without losing who you are at the core.

The answer? You smartly adapt, starting with the less conspicuous ones. My retired professor friend stated that his old country's culture practiced the wife/ mom as the preparer of family meals. Such an arrangement was sustainable in a culture where the husband as the breadwinner toiled outside the home. The wife/ mother tended the affairs of the family at home, including meal preparation. When he and his wife moved to the United States, his wife picked up work at a nearby organization. He sometimes came home before his wife. He practically would prepare meals for both he and his wife to eat when his wife returned home. After all, he reasoned, no one would die if he performed this minor role reversal. At first, his wife was pleasantly shocked and became comfortable with this arrangement in their new land. Then, something interesting occurred on his wife's part.

In their erstwhile culture, the husband as the breadwinner provided all of the family needs financially, including the payment of all household bills and sundry purchases. Since his wife was now working and earning a living, he approached her to partake of the family's financial obligations, no matter how minor that was. Upon hearing this and still stuck in the old country mentality, his wife did not take the suggestion well. You see, in the old country's culture, such a proposal by a husband would be deemed dereliction of his responsibility as a caring and provider husband. It would demonstrate weakness, laziness, and lack of fortitude in the old country. After some days of reflection and pondering, his wife realized that it was a new day in a new land. It became abundantly clear to the professor's wife and, of course, the professor that some adjustment to their mindset on certain items was inevitable.

Aisha and her husband Ahmed had emigrated from the middle eastern region of the world. In their old country, women played specific roles and would not cross to what society regarded as male roles. One of those prohibited roles for women, as defined by society, was automotive driving. While they lived in their old country, Aisha utilized a male chauffeur who was paid monthly as his regular job. The driver took Aisha wherever she went most of the day. Then Aisha and Ahmed moved to the United States. Here, chauffeurs do not come easy. They are available but at a very steep price. The couple could not afford such a luxury here in America. Two months after their arrival to the United States, Ahmed secured a job with his electrical engineering degree and credentials. He utilized the only car they had at that time to go to work. Aisha was stuck in the house and could not even go to the store while Ahmed was at work. She could not drive. She never learned how to drive in the old country. How could she, when it was not permissible? Frustration was mounting for the young couple. Aisha was depressed being confined to the house, and Ahmed was exhausted from playing a chauffeur to his wife upon return from his demanding job. Something had to give.

The husband and wife realized that they had to adjust to the reality of their new land. They agreed for Aisha to learn to drive at a nearby school. Within two months, Aisha was proficient at driving. They then purchased a pre-owned car for Aisha's use. Aisha was grinning from ear to ear the day they bought and brought their second car home. She quietly called her mother in her old country and joyfully announced that they bought another car she would be driving. Aisha's mother, still in the old country in a muffled voice, congratulated them but advised them to keep their affairs quiet since it was still prohibited for women to drive there. Of course, Aisha and Ahmed could care less about the rules in the old country. They no longer belonged there and happily so. Aisha and Ahmed have morphed into their new hybrid selves.

Oh, not yet done with Aisha and Ahmed. The trouser! The pants!! What about? You may ask. It was such a big deal as Aisha's driving, or perhaps even bigger. You be the judge. Aisha had never donned a pair of pants in her life. She had only dreamed of wearing one. Aisha, like most Millennial cohorts around the world, is steeped in technology and digital devices. So, Aisha was a prolific user of social media and the likes. She owned and operated accounts on Facebook, Instagram, WhatsApp, and Twitter.

Consequently, she kept pace with global social events and celebrities. She could tell you Lady Gaga's comments on Twitter and what Rihanna wore to a televised event or posted on her Instagram account. She craved and yearned for these modern niceties and styles, just that she could not physically indulge in them in her old country. Females were prohibited from wearing pants or shorts in the old country. Violation of such rules attracted fines and punishment. Violators were looked down upon as non-conformists and rebellious. A provocative dressing was regarded as flirtatiousness and equated to waywardness and a lack of decorum. Parents and the extended family exerted tremendous pressure on children, especially females, to strictly conform to the societal norms.

With wider technological and digital proliferation, the world has become smaller, such that cultures, attitudes, and behaviors are shared instantly and expansively. Aisha was into the western and modern ways but had suppressed such tendencies while in the old country, at least publicly. Her husband, who was well aware of Aisha's western inclinations, would often warn her to be careful not to attract the community norm police. Aisha heeded to such admonishment and bottled her feelings discreetly. You could then imagine Aisha's behavior and attitude once she and her husband emigrated to the United States. She exploded in excitement. After arrival and settling in their new residence for about a week, Aisha sought and identified the nearby clothing stores in a shopping mall three miles from their home.

Ahmed knew it was coming and waited for it. During the early hours of a Saturday, Aisha prepared breakfast, making her husband's favorite cuisine, an omelet with all the ingredients and condiments her husband would die for. Then, during breakfast, Aisha popped the question if her husband would take her to the mall for a day of shopping. Poor Ahmed, who had just been bribed with a delicious breakfast, could not possibly turn her down. Within two hours, they were in the mall. Aisha was giddy and visibly excited. She led the way to a ladies' clothing store while Ahmed followed hopelessly and helplessly. Aisha dashed straight to the pants section. She knew what types of pants she wanted. After all, she had already worn them in her dreams and imaginations all these years in their old country. She grabbed a handful of jeans pants and a couple of regular pants. Ahmed panicked and bellowed, "Oh my God, that many?" At which time, Aisha explained that she was going to try these pants and pick only two. Ahmed gave a big sigh of relief.

For Aisha, this was more than purchasing some pairs of pants. This experience represented a significant aspect of her life's aspiration and dream. Now, she was living it. One by one, she wore every pair of pants she took into the dressing rooms. She would walk out after wearing each pair with a reassuring stride as if to say: see me now. She would then walk to the giant panoramic viewing mirrors, flex, and ask Ahmed his opinion on each try out. With mutual evaluations, they settled on a couple of pairs. Some of us who began wearing pants from infancy may seem perplexed how purchasing and wearing a pair of pants can be that dramatic and life-changing. The simple truth was that it was mammoth for Aisha. It was about her new self and identity made possible by their emigration to a new land, America, where the freedoms and ways of life allowed her and Ahmed to exhale.

I will not bore you with the added story of the first time Aisha wore her newly acquired pair of pants to an event. Ahmed teased that she did not sleep well the night before because she was anxiously watching the clock most of the night. It was as though she was counting down on the launch of her new pair of pants. Today, Aisha possesses numerous pairs of pants, and her initial heightened excitement about wearing pants has waned. Interestingly, according to Ahmed, she still kept those first two pairs of pants as mementos. She could not get into them anyway, even if she tried to wear them. Since she initially purchased the pants, Aisha has had three children and packed additional twenty pounds over the years. She would never get rid of those two pairs of pants, Ahmed predicted. They represent her new life, her emancipation from the shackles of a society that suppressed her yearnings: the gaining of her social freedom, and her "outward exhalation."

Morris had emigrated from a tropical climate region to attend college. He fell in love with an American classmate at medical school. They got married and settled in Seattle, Washington. Before emigrating to the United States, Morris never physically experienced snow or snowfall. The closest that he came to snow was on TV and the movies. Morris's wife, Stephanie, on the other hand, was born and raised in the Northwestern United States. Stephanie and her family were avid skiers. Stephanie's older brother represented the United States in an Olympic skiing competition in 1988 at the winter games in Calgary, Alberta, Canada. Stephanie could comfortably live on a ski slope. As Morris and Stephanie dated, Stephanie exposed Morris to skiing. Morris was very apprehensive at first and felt like a fish out of water. The concept of rolling down some white wet sand-looking slope, often endangering oneself for the sake of some exhilaration, seemed alien to Morris. Initially, they would go to the slopes, and Morris would confine himself to the clubhouse and drink chocolate, his favorite winter warmer. Morris was a bright and competitive individual who was an achiever with a strong drive. After watching the skiers from the clubhouse's warmth numerous times, he decided to venture to the slopes. You see, Stephanie knew her husband very well. She knew he would one day engage in the sport, on his own volition. She laid low because she reasoned that if he plunged on his own accord, he could not abandon the effort without succeeding. That was her husband. So, she never pushed or even suggested his taking up the sport. Stephanie's genius proved true.

One day, as they drove home from the ski resort, Morris opened the conversation by asking Stephanie what equipment he needed for a starter skier. Stephanie controlled her joy by casually and unemotionally mentioning the required basic equipment. When Morris suggested they purchase these items for his trial of the sport, Stephanie reeled him in by offering they obtain some used equipment from their parent's home, where tons of ski equipment and paraphernalia abound. Stephanie posited to Morris that by taking that approach if it did not work out, they would simply return the items. All along, Stephanie knew that Morris never quits whatever endeavor he pursues. She was, therefore, just playing along nonchalantly. Morris took the bait. The following week they were at Stephanie's parent's home rummaging through a heap of ski equipment. They found what they needed.

Two weeks later, they were at the slope with Morris, guided and trained by a professional they engaged for three months. By the third month, Morris could ski down the hill at his speed. Morris finally caught the ski fever and bug. They never looked back. They are now avid skiers who have skied in the country's most popular ski spots, including Colorado's Aspen, Vail, Breckenridge, and Steamboat, Utah's Park City, Wyoming's Jackson Hole, and Idaho's Sun Valley, to name a few. Morris had morphed into a hybrid individual incorporating some culture, habits, and inclinations of his new land.

A phenomenon that occurs with some newcomers needs a mention here. It is about the modification of one's name for cultural and societal fit. Sometimes, this takes on a complete name change. Many who have engaged in this adaptation cite their given long names or family names. Others have blamed the challenging pronunciation of their name due to its origin and phonetics difficulties. Whatever the reasons are, we remain nonjudgmental, but it points to the ardent expectation and pressure to conform in a new land. Some cling to their given names no matter how it is uttered or spelled. They become numb to its massacre, choosing to stick with its cultural meaning and purpose than the need to fit in.

In the early eighties, a schoolmate who had emigrated to the United States picked a new name three years after his arrival. He lived in a sprawling apartment complex with over four hundred units. We talked, and he invited me to his place. Remember, this was an era when mobile phones were non-existent, and "Siri" and the Global Positioning System (GPS) were only conceptions in the minds of scientists. I managed to find my way to their massive complex and his stated building. I thought that I was smarter than I really was and did not write down his exact apartment number. As I got to his building, I forgot the number he told me and was staring at many apartments with no idea how to proceed. Again, no cell phones then but only landlines, and I could not ask a stranger for the use of his or her phone inside their apartment. First of all, for safety and security reasons, it could have been challenging to be allowed access into a total stranger's house to use the phone. Compounding the matter was that the neighborhood was not Rodeo Drive in Los Angeles or Magnificent Mile in Chicago. It was a semi-run down area. I resorted to hanging around the building and asking anyone entering or exiting the building if they knew which apartment my schoolmate lived in. Nobody had heard of the names I muttered; no luck whatsoever.

In frustration, I began heading towards my car parked along the street. At that moment, I heard my name shouted from across the road and my schoolmate walking briskly towards me. He had waited for a knock at the door. Upon guessing that something was amiss, he came out of his apartment to look around. He jokingly asked what my problem was. Couldn't I read the apartment numbers displayed at the door? It was then that I shared that I did not write it down, believing I would recall it from memory, but my memory failed me. I quickly tore into him by questioning how no one in their building knew who he was. He calmly inquired what name(s) I used. I snapped back by asking what that line of questioning meant. He immediately broke out laughing. By this time, we had made it inside his apartment and seated. Then he began, "You know how long my native name is and whereas it was easy for it to be pronounced at home, it became a nightmare here. So, with the blessing of my family, I changed my name to Elvis. I now go by Elvis, sorry I forgot to mention that to you the few times that we spoke." I could not wrap my head around it at that moment. How can one change their given name so that others can pronounce it right? I murmured under my breath. Unbelievable! No, not just incredible. It was plain ridiculous, I reasoned. It took me some years to understand it, though I still have my reservations about such a draconian move, in my opinion. Yes, adaptability, assimilation, and acculturation come in different shapes and flavors.

How about this food story told to me by the wife of a friend? She had a childhood friend who emigrated before her to the United States and lived in Boston. They attended the same high school in their old country and were quite close. My friend and his wife went on a vacation in and around Boston. They invited her friend to join them for dinner one evening. Since she lived in Boston, they entrusted her to select a good restaurant for a nice dinner. Her friend quickly assured the couple that she knew of a few nice eating establishments.

So, they gathered at a posh restaurant near the waterfront in Boston. It was a lovely location with a comfortable ambiance. The menu reflected the eliteness of the establishment. Then, it came time to order food. As a courtesy, her friend signaled them to place their orders first, and they did. It was her friend's turn now. Her friend ordered assorted sushi, three different types to be exact, with accompanying wasabi sauce to boot. My friend's wife recounted that she was shocked at first then her shock turned to curiosity, which she claimed settled to some confusion and slight disgust. What was going on here? She asked herself. Her friend eating raw fish, raw uncooked fish? Unbelievable!

When their orders were delivered, her friend delved into her food and seemed to be enjoying her entrée. My friend's wife ate her food and partially enjoyed it as she was fixated on her friend's new palate of "disgusting raw fish" as she termed it. After the main dishes, they ordered some desserts. It was during dessert that my friend's wife could not hold it any longer. She must find out what was happening. So, she began, "Grace (her friend's name), we noticed that you ordered and ate raw fish for dinner and what was more, you seemed to have enjoyed it. What are you trying to prove? That you can consume raw things?" Grace was very composed and started by saying that she was surprised her commentary waited until then. She expected their curiosity and comments. Grace explained that several years ago, she dated a classmate and fellow emigre from Asia. During their relationship, he exposed her to sushi "At first," she continued, "I could not stand the smell of raw fish what more the taste. I rejected it outright. But after some time, I indulged in its consumption and eventually love it tremendously now. It is an acquired taste," she explained, "that one builds on overtime." She concluded by stating, "I do not blame your initial aversion towards it. I felt that way at my first exposure to it and by the way, not all sushi is made with raw fish. Some contain cooked fish." My friend's wife and her husband listened but could not understand it, including Grace's rationalizations, or chose not to understand it. They felt it was a significant turn for Grace. She remembered that Grace would not eat anything, fish, meat, or poultry with any tint of blood in the old country, and now she is gulping raw fish down her throat. They concluded that she was a confused person that was embracing anything thrown at her. Pitiful, they judged.

Most newcomers who emigrated from rural areas are generally closer to nature. This intimacy is because they resided in environments surrounded by nature, such as mountains, hills, rivers, and lakes. These newcomers tend to be attracted to the urban and metropolitan centers upon arrival. So, it was surprising to observe Lucas, who emigrated from a semi-rural region of Brazil. Lucas became an avid hiker in the United States, his new country. He did not hike at all in his old country. As he explained, "Our daily lives were one big hiking exercise. Everywhere we went, we trekked." Why would Lucas re-engage in what he had performed all of his life? What was the thrill for him after all?

Let us take a moment to explore what hiking is. The dictionary definition of hiking is walking in nature as a recreational activity. A natural exercise that promotes physical fitness. Others have described hiking as immersion in the outdoors using one's feet as transportation, at one's pace, and carrying whatever items needed for the day to discover the beauty of nature. Yet others have termed it as nature's therapy that is affordable and plentiful. It soothes and clears the mind, rejuvenates, and elevates overall well-being through its beneficial effect on our mental health. This understanding and appreciation are where Lucas's story became interesting.

You see, one would have expected Lucas to be averse to nature and her elements because of the heavy exposure to it his entire life in his old country. But that meant viewing hiking less wholesomely. As defined and described in the preceding paragraph, hiking goes beyond walking the trails. It is an elixir for the mind and soul. It provides nourishment and the avenue to connect with nature and its splendor. Lucas understood that and leveraged every bit of the opportunity. Whereas Lucas did not need to hike in his new land for sustenance, he combined his appreciation for nature with a quest for mental and perhaps spiritual calmness and meaning. By doing so, Lucas transformed into a hybrid Lucas.

In a new land, one is susceptible to the influences of different and varying cultures. The adoption and sometimes synthesis of these cultures produce a modified self. You become a hybrid self.

"Strength lies in differences, not similarities."
-Stephen R. Covey

CHAPTER 17

LOOK FOR THE GEM
AND HARNESS!

"He had all of the essential skills, especially the technical competencies, but his English language is not up to par." These were the exact words of a midsize manufacturing plant's department hiring manager in the United States' midwestern region. He continued by saying, "I settled for another candidate, whom though not as technically competent, spoke good English, heck, he was born here." This hiring manager just threw away a gem handed to him. He was myopic in this particular hiring approach and decision-making. You will be surprised that this narrow focus is more prevalent in the workplace and society than imagined. Our unconscious biases and tunnel vision limit our ability to fully see and pluck the roses from the bushes pertaining newcomers to our nation. It is often not intentional or even deliberate, rather our knee-jerk reactions and attitudes towards new entrants to the land and the workplace.

It has been reported in the literature that American attitudes and treatment of newcomers are both ambiguous and ambivalent. The effort and intentionality required in enabling newcomers to integrate into the mainstream successfully are, for the most part, taken for granted. In fairness to leaders, hiring managers, and human resources professionals, very little appears to be available to them in effectively integrating foreign-born workers in the American workplace. Perhaps, it has not been deemed a vital component of our workplace ecosystem. It is now gaining more attention, considering the workplace and general population dynamics. We must confront and address this growing necessity and emerging competency in the work and societal environments. Until now, it was easier to seek the "already-made" candidates and impatient in digging deeper to extract the best in the long run with newcomers. One has to be culturally intelligent to navigate and apply the skills in this sphere successfully.

David Livermore observed in his book *"Leading with Cultural Intelligence,"* that cultural intelligence is a critical capacity for living and functioning in today's world. One must master the art of fast understanding, adapting, and applying our different cultural backgrounds and settings. However, the social, cultural, educational, and psychological components in the newcomer's milieu are complex and taxing. Before you accuse me of wanton generalizations and incorrect assumptions, could you please pause for a moment and ponder if you or someone you know or work with has ever done one or more of the following:

Rationalized not hiring someone (newcomer) because of the little or no proficiency in the English language? Even though the actual task entails seclusion at a corner of the organization or monotonous and robotic interaction with machinery and equipment. After all, lack of proficiency in the English language is not a permanent personal attribute. Would a little investment in language training have helped, especially if other skills are identified as reliable? Have you or someone you know or work with advanced that a candidate seemed too tepid or reserved to land the position with high marks on all other measurements? Could it have been the result of the culture from whence the newcomer just emerged? Could this docile talent bloom later when the integration and assimilation into the culture have fully occurred? Have you or someone you know or work with been dismayed by the inability of a new entrant to fully comprehend your statements, including slang and vernacular, or not able to articulate their thoughts, ideas, and views properly? Even though the interactor's language pattern and style are replete with colloquialisms and local vernacular. Or better yet, wondering dismissively why the newcomer had not taken time to learn English before emigrating to the United States? After all, he or she knew that the language of the U.S. is English.

Let us extend this a bit. Have you or someone you know or work with insisted or argued that the organization must not spend its limited budget on English language training for newcomers or translating company-wide correspondence into another language even if a sizable percentage of the workforce comprises of this ethnic community? Have you or someone you know or work with not understand why the newcomer cannot just eat the main dishes and offerings at the company outing or party? After all, they would not perish if they ate it only this one time; the person would reason. How about getting irritated that the newcomer would not sit with all and mingle during the company lunch hour? Have you or someone you know or work with complained about an employee who takes specific days off yearly for religious activities?

Again, I have insisted throughout this book that these questions are not intended to indict or laden the reader with guilt. Instead, they are meant to arouse and perhaps cause us to examine some unconscious feelings and attitudes to make any course correction if needed. Furthermore, they are to wear the newcomer's shoes even for a split second. Such an exercise makes for a better human understanding and community. We all have what has been referred to in the literature as cultural baggage. This encumbrance comprises our beliefs, values, biases, prejudices, and habits. Our cultural baggage makes us exhibit certain feelings and emotions, such as joy, fear, discomfort, and frustration when exposed to cultures different from ours. Often, it is our cultural background that evokes these sentiments, not the other person's culture. This observation is because we can only control our emotions and not impact how someone else was raised or their background makeup. Also, I must reiterate that for all intents and purposes, the newcomer referred to at any time and place in this book is a legal newcomer who emigrated to this nation with all required U.S. government approvals.

One fundamental assertion that could be made of newcomers who endured the arduous ordeal of conceiving the idea, seeking the path, and engaging all required government agencies of the country of birth and the emigrating nation is that they are persevering and tenacious. The migration process is not only tasking but tests one's mettle. I discussed these in the earlier chapters of this book. One must pack a lot of patience and fortitude to withstand the rigors and consternation associated with migration. So, what has that got to do with the here and now? After all, the newcomer is now in his or her final destination, America. The answer is that lots are gleaned from this baptism of fire. Resilience! If resiliency, perseverance, and tenacity are attributes required of a position, the hiring manager might check these items off as accomplished. There should be no need to administer

EMEKA OKEANI

a psychological test to cull out this attribute. It exists and screams with the newcomers.

To fully harness the new entrant's skills and aptitudes, the hiring manager or supervisor must be culturally aware. Let me go one step further and state that the hiring person must be culturally intelligent. What is this lexicon, one may wonder? It represents the ability to adeptly and seamlessly recognize, appreciate, and empathize with various cultural contexts, situations, and environments yet not losing one's core culture. Cultural intelligence is a learned tool, regardless of one's exposure. In other words, it is an individual capability that is brought alive by intentionality and effort. It is sometimes referred to as intercultural fluency. One can live amid a different ethnic community and be oblivious of the cultural sensibilities that sustain such a group. This lack of awareness and understanding occurs when either artificial or real cultural blinders are erected. I venture to say that cultural blindness creeps upon us or is stealthy at best. Even those that regard themselves as culturally aware and sophisticated may find themselves lacking sometimes. Organizations also fall prey to this deficiency in cultural agility.

ABC Company in the Northeastern United States maintained and utilized an interview template for all positions within the organization. Some of the questions and expected answers were designed to gauge team orientation. One's ability to work comfortably in a team versus solo work. What if the candidate's old country and work culture required individual effort and interpersonal competition? Examples of teamwork and cohesion may be scarce for the candidate to relay. Their cultural background usually skews the candidate's overall experience. It is not advocated for ABC company to compromise or lower its hiring standards. Instead, hiring managers should factor in and weigh cultural influences and backgrounds realistically. In this example, effort should be made to determine the candidate's ability and fluidity in transitioning from solo work to team orientation. Another often encountered example is the emphasis on directness in communication within the work setting. Interview questions seek to determine the candidate's ability on assertiveness and personal confidence. Directness is an American value and behavior embedded in her culture.

Many cultures worldwide take an opposite approach to directness, purporting such to represent rudeness and disrespectfulness. Therefore, should a candidate, a newcomer, lack this virtue and excel in most other aspects in an interview, does the hiring manager cut the candidate some slack, factoring in the cultural background whence the newcomer came from? It is very tempting for us to assume that a hiring manager weighs these items thoroughly. It is easier said and intellectualized than done. We often unintentionally allow these filters to cloud our decisions.

Mary was the department head of an IT firm specializing in marketing data consolidations and mining. They needed to fill some positions in their rapidly growing company. Sue applied and made it to the final interview. The final interview was a panel interview comprising three managers in the department, including Mary. Sue impressed all three interviewing managers on all aspects except her seeming "coldness," they concluded. They chose another candidate ultimately. You see, Sue's supposed timidity could have been culturally induced. Chances were that in Sue's old country's culture, meekness and respect trumped directness and boldness. Unfortunately, in some cultures, such as in the United States, such a docile disposition could be interpreted as weakness, tepidity, or plain lack of confidence. Not necessarily so! The point here is that Mary and her compatriots should have been more culturally aware in their assessments and final decision-making. Their company may have let a gem slip through due to their intercultural fluency deficit. The point that needs to be continuously made is that these cultural failings are not blatant and glaring. They are often hidden, subtle, and unsuspecting. A hiring manager must always remain vigilant to avoid any ambush in this area.

The onus of cultural awareness lies on the newcomer as well. The newcomer must quickly ramp up their knowledge and skills. For example, a male newcomer who hails from a society that is unaccepting and disrespectful of female bosses must understand and accept those female bosses represent an embedded element of the American workplace terrain. However, the newcomer is still "new" to the American culture, attitudes, and habits. Over time, the newcomer adjusts and assimilates adequately. In the meantime, the preponderance is on the native hiring manager to step up regarding cultural awareness and fluency. For example, when a hiring manager asks about a newcomer candidate's experience in a particular area, the newcomer will pull from the old country's experiences. Sometimes, the experiences or expectations may not produce the American equivalence. Here then lies the needed unscrambling or interpretation of the newcomer's told examples into an American match for accurate evaluation. To perform this matching properly, the hiring manager must be interculturally fluent. Otherwise, valuable information on past experiences of newcomers is lost in cultural translations and interpretations. These misreads could deny organizations candidate gems that could contribute immensely towards their mission and goals.

Sometimes, the best newcomer candidates are missed due to inadequate training and the hiring managers' lack of sophistication. Consider a hiring manager who is well-intentioned but lacks the skills to vet newcomers properly. The hiring manager may be great at evaluating native-born candidates but inexperienced when handling newcomers with their inevitable uniqueness and peculiarities. A colleague told the story of a respected hiring manager in a relatively small supply company who interviewed a candidate who just emigrated from Eastern Europe. After the candidate drove off from the company's parking lot, the hiring manager came into the breakroom where her colleagues were having a coffee break and announced, "I just interviewed a candidate from Slovakia with an 'interesting accent'." A peer asked, "How do you know where she came from? The hiring manager proudly said I asked her. "I told her right off the bat that I hear an accent, and asked where is she from" Ouch! And Ouch again!!

Whereas the hiring manager may just be curious, the line of questioning was entirely inappropriate. Also, imagine a truckload of unconscious bias that must be hovering around that interview room. Besides, imagine perhaps how put-off the candidate may be initiating an interview with her country of origin. She just left there and for a reason. You do not want to take her back to such a place in an interview. What if she was fleeing from some form of persecution- religious or political? Interviews are mutual assessments. The newcomer may decide to decline any offers due to what she perceived as cultural insensitivity. After all, the hiring manager at that moment represents the company. Sometimes the role is reversed. Consider when an interviewee was provided the opportunity to ask questions of his four-panel interviewers. His first question occurred as he turned towards the Human

Resources Manager, who emigrated from South America, with some slight accent, and questioned, "Where are you originally from?" The Human Resources Manager recounted that the four interviewers froze by such insensitive and unwarranted question. A job candidate seemingly fixated on determining an interviewee's country of origin – bad move. Needless to say, he did not land the position.

Some culturally insensitive and inappropriate interview questions that I have heard from people told over the years include the following: "How was your childhood?" Really? Maybe the inquirer wanted to know if the candidate would need therapeutic intervention for early childhood abuse? Please pardon me if I drifted into sarcasm. I am human, you know. Here is another one, and you be the judge of it. "Do you think you could work for a female boss?" Or how about this one "Do you believe in fate?" Some interviewers may sneakily ask, "What is your background?" hoping the candidate voluntarily offers information on their ethnicity and national origin. "What country are your parents from?" believing it is permissible since it is not directly about the candidate. Often, these questions seem well-meaning out of curiosity, belief that it was necessary, or dabbling into what they consider a friendly conversation. These questions, however, end up being counter-productive and, in some cases, hurtful and demeaning. You could miss your newcomer candidate roses!

The way and manner you read the newcomer's body language may cause you to miss the mark. This miscalculation is because body positions, moves, gestures, and interpretations are different in various cultures. When interviewing or engaging with a newcomer, you must consciously sift through your interpretations and observations. For example, with eye contact, the American culture encourages direct eye contact as a sign of strong engagement with your subject. In some cultures, direct eye contact is deemed disrespectful and plain defiance in some cases. So, as a hiring manager, you must not interpret eye contact avoidance as disinterest or flat-out timidity. In some cultures, in Africa and some other places, eye contact is graduated with age. One must not directly look at an elder, but it is permissible for peers to look at each other directly. In other cultures, it is even gender-filtered. Males can directly look at each other while females are not afforded such a privilege. Let us stick with the eye a little longer. Winking remains inappropriate in some cultures, while some form of winking is acceptable, such as American. For example, when you wink, say to your child or younger person, especially in the performance of an act, or task, or competitive event, that signals, "Atta boy!" which represents encouragement and positive affirmation. The same winking act, especially to a member of the opposite sex, would be deemed inappropriate and flirtatious.

The hand occupies a vital spot in the discourse of body language. Some hand gestures are universal, such as waving goodbye or beckoning someone with your hand. Hand clapping is also ubiquitous and harmless. The hand gestures and finger manipulations go into a different realm of meanings and interpretations. We will not dissect and examine the numerous finger gestures within different cultures. I discussed some of those in chapter 14. We will need an entire chapter to do justice to such a construct. The one area that is worth visiting is the old age handshake. In the American cultural context, a firm handshake is preferred over a soft or sweaty one. In America, a firm handshake depicts confidence, self-assuredness, and a positive attitude. It is such a valued and regarded behavior that most people, especially in the business and political world, try to outdo each other. If you want to test my hypothesis on this, watch political candidates square off for a debate and observe their often-practiced handshake of their opponent. Some consultants make a living teaching the art of handshaking and body language in general. In some cultures, a weak handshake is preferred or at least neutral in perception. A firm handshake is regarded as aggressiveness and combativeness and therefore discouraged. For the hiring manager, consider these cultural variations when evaluating the newcomer, so you do not miss a gem.

Let us discuss space and touch. In America, you must provide one their personal space, otherwise referred to as elbow room. Any incursion in one's space is frowned upon. In some cultures, getting closer to someone space-wise signifies comfort with that individual. The cousin of space is touch. There remain basic understandings and expectations with touch in all cultures. The newcomer may still be at a learning stage regarding these behaviors and practices. Therefore, understanding and appreciating the newcomer's level of acculturation is essential. May I emphasize that these suggestions are not giving the newcomer a "free ride."? The newcomer must also be responsible for quickly learning the major dos and don'ts culturally in their new land. What is espoused here is not letting a few missteps occasioned by the newcomers' newness obscure the pent-up talent and skill. Go beyond the peripheral cobwebs and seek the hidden competencies ready to be unleashed.

Be aware that newcomers typically opt for or pursue jobs and tasks that are less than their responsibilities in their old countries. This approach is usually taken for two reasons; easing into the American job market and needing time to match their skills and competencies with the U.S.

equivalences properly. As a result, the newcomers often under-sell themselves. As an adept hiring manager, you must dig deeper to adequately extract and assess the newcomers' true talents and skills. In other words, go in prepared to explore, and you might come away with a boatload of goodies.

Unconscious biases, which represent social attitudes about certain groups of people outside our conscious awareness, are prevalent in the workplace, especially in hiring. It is that set of stereotypes that people unconsciously attribute to another person or group of people that influence how they perceive and interact with that individual or group. Unconscious bias is a stealth operator in that it is often incompatible with our conscious values. Most people are superficially cognizant of this subtle and opaque thief. It is usually implicit and under the radar. Behaviorists tell us that certain scenarios and circumstances such as heightened stress trigger unconscious attitudes and behaviors. Through research, some examples of unconscious bias include white-sounding resumes receiving more callbacks for interviews, healthcare professionals, and providers' decision-making hinging, to some extent, on bias instead of strictly science, and the selection of doctors within the medical practitioners' insurance pool.

A business acquaintance who emigrated from Japan a few decades ago told me the following story. He successfully climbed his organization's career ladder to become the Chief Executive Officer of their company. As a practical and down-to-earth executive, he indulged in covering some chores reserved for the rank-and-file employees. For example, he would answer their company's main phone if the receptionist was not at her desk. He did not mind and viewed such behavior as the versatility of tasks to all employees. One late afternoon, he was reading and reviewing some correspondence when a call came into the company's main phone line. He answered with the company's name. The caller requested to speak with the CEO.

That was the caller's first attempt at reaching the CEO. The CEO gladly intoned, "This is he." There was a short pregnant pause at the caller's end. Then the caller resumed, "I don't think you understood me. I want to speak to the CEO, please." The CEO repeated, "That's me." The caller got irritated and showed it by yelling, "Well, I just have to call back tomorrow because you are not understanding my request," and hung up. You see, though the CEO had lived in the United States for over three decades, he still possessed his distinct accent, which was not particularly American, any part of American accent or dialect. The caller had surmised that one with such an accent could not possibly have been the CEO. The caller eventually met physically with the CEO after some weeks. The caller embarrassingly confessed that he did not think that one with an accent (implying non-native-born) could have risen to that national powerhouse's CEO position. This story reflects an unconscious bias on steroids. It happens.

During my undergraduate days, I engaged in an "experiment" in our sociology class with two other classmates. It was a project on perceptions. In the early eighties, unconscious bias concepts and lingo had not permeated social psychology literature. At least to the extent that it has now. In retrospect, our project's more appropriate title would have been "Unconscious Bias." We were fumbling and bumbling into unconscious bias without even realizing it. There were three of us, me with African ethnicity, Sam with Eastern Indian heritage, and Brad who was born and raised in Nebraska. Brad is Caucasian. The class assignment was to identify any medium in a society where one's perception could be clouded or skewed by visual images. After some deliberations, we chose fast food drive-through scenes. Sam had hypothesized that in fast food drive-throughs, whenever the order taker saw him in his car, it took a long time to order because the order taker would claim not understanding him, and he would repeat his order numerous times. But for the drive-throughs where the order taker did not see the customer, the order was accurate without hesitation. We all agreed that it represented an excellent subject to explore.

The observations and findings blew our minds. However, it does not constitute a scientific experiment. Our sample size was grossly small, and we did not institute all the required validity rigors of an actual scientific experiment. Nonetheless, the directional nature of the results was eye-opening. For over ninety-eight percent of the orders placed by any one of the three of us through the unseen customer, orders were understood and taken instantly without request for a repeat for not understanding the customer. Conversely, for the drive-throughs where the customer is seen, Sam and I recorded an eighty percent repeat request, while our Nebraska born and raised classmate Brad had zero, yes, zero repeat request. We may have influenced our "experiment" in that Sam and I wore our ethnic native attires during our experimentation, while Brad wore regular jeans and a casual shirt.

One may throw a shade here by submitting that well; Sam and I genuinely possessed some distinct accent that may have been hard to decipher. Oh yes, and that was precisely the point. Such accents would have caused the order taker to have requested a repeat of our orders when not seen. Instead, we breezed through our order without any claim of not understanding us. To perhaps lend some amateurish validity to our "experiment," whenever we went through the drive-through and obtained our result, we parked our car and went inside to place another order. We received an equally high percentage, nearly eighty-two, of repeat order request when we showed up in front of the counter and our native regalia. However, we must mention that the drive-through order taker was usually not the same worker that took our order inside. The incidence of ordering from the same drive-through and counter order taker was less than seven percent. The point highlighted in this story and "experiment" is that unconscious biases follow us in our daily lives. The young order takers at the fast food restaurants did not mean any harm at all. They most likely were not even aware that unconscious bias was their "buddies" during these times. And that is the danger of unconscious bias. It is covert and mostly undetectable. But it lurks, and all the time.

Let us examine one more example of this surreptitious raider of our brain, unconscious bias. Doctor "A" was born in this country of second-generation immigrant parents. He attended a reputable medical school and is board-certified as a primary care physician. For many years doctor "A" worked for a regional healthcare system. He performed his duties diligently and received good ratings from his patients and employer alike. He resigned from his healthcare network position and opened a private practice. Things did not take off as he had imagined. Patient traffic was meager, and referrals were minimal. A fellow physician friend suggested engaging a local marketing company's services to assist in his growth plan. He obliged and signed up a market research outfit accordingly.

After two months, the company met with doctor "A" to present their research, analysis, and recommendations. The marketing company provided many valuable and practical suggestions. The one revelation that numbed doctor "A" was the insinuation that his name, last name, to be precise, might be affecting his selection within the insurance pool. According to the company, the doctor's last name sounded "foreign." They explained that patients stick with the familiar and comfortable regarding personal health matters and decisions. Though the patients know that listed doctors are board-certified, they wonder if the doctor was trained in the United States, end to end. That slight unconscious apprehension compels them to bypass a doctor with such names and select a doctor they feel received all of their education and training in America.

Remember that doctor "A" was not only born here but also obtained his training in the United States but at this stage, it mattered less for the deciding patient. Unconscious bias is extremely fleeting and moves at a rocket speed. Before the patient thinks about their decision process, this creepy element does its damage by impacting a decision. Sometimes, the victim might not even recognize what transpired. No one faults a rational decision-making process. The clouding of this process by unconscious bias remains the concern.

The saving grace? Unconscious biases scientists tell us are not permanent. They are malleable, and intervention and steps can be taken to mitigate their impact on our thoughts and behaviors.

Confirmation bias represents one of the various aspects of unconscious biases. Avoiding confirmation bias is essential in hiring assessments and conclusions. This bias occurs when you review a resume and form an initial opinion of the candidate based on attributes that do not directly reflect the person's skills, experiences, and work ethic. These include but are not limited to name, ethnicity, school attended, and so forth. An opinion formed at this initial stage invariably drags into the actual interview session, thereby allowing the insignificant, inconsequential views to cloud an unbiased assessment of the candidate. The proven way to keeping this sneaky bias in check is by steering interview questions towards standardized skill-based ones that would neutralize any conceived confirmation bias.

The hiring manager must avoid or at the least control what has been dubbed attribution bias. This feeling refers to when a hiring manager assumes and concludes on a hunch without proverbially peeling the onions. For example, if an interview candidate is late to the interview, instead of writing them off as unfit for the position due to lateness, ask what happened. It could be an acceptable reason or incident. Furthermore, be aware that an interviewing process triggers nervousness, distractedness, and plain mental block. Dig deeper and gun for full stories before judging.

As humans who represent a social and environmental species, we garner feelings, attitudes, beliefs, and opinions as we live our lives. We inevitably carry positive and negative biases on matters and items. We are inertly wired to sort and organize things. What is being emphasized here is that we become acutely aware of these pouches of biases and ensure we minimize their errant and unfiltered influences. For example, many unconscious biases follow us around gender, age, height, beauty, name, and ethnicity. We must work at shooing these away in our decision-making process.

This discussion takes me to an example of harnessing the gem. In the late eighties and early nineties, a boss was highly adept at picking the best talents from a heap of candidates. What was extraordinary about his skill rested on his capacity to decipher talent among newcomers. He would seamlessly wade through any bias and distraction and select candidates who excelled at their responsibilities. His batting average on his candidate decisions was near perfect. It did not matter what part of the world the new entrant came from, Africa, Asia, Europe, Micronesia, etc. I always wondered what made him uniquely competent at this skill. I unscientifically concluded that it must be attributable to his worldwide exposure. You see, my boss, who was born and raised in the United States, spent two decades in the military traveling and living in many parts of the world. Therefore, he was exposed to varying cultures and ways of life. He had become culturally intelligent. Armed with this powerful tool, he never allowed cultural blinders and biases to permeate his decision-making, especially with job candidates. He was, therefore, able to home in on the talent before him and pluck the gem.

Traveling outside one's comfort zone and geographical location improves understanding of different cultures. Exposure and awareness mitigate implicit biases. Fear fostered by lack of knowledge breeds intolerance and misunderstanding. Also, it is posited that experiencing specific challenges in another cultural setting or society enhances our capacity to empathize with newcomers to our land. Philosopher Saint Augustine stated, "The world is a book, and those who do not travel read only one page." A century ago, renowned American writer Mark Twain expounded, "Travel is fatal to prejudice, bigotry, and narrow-mindedness." I further assert that based on my interactions with individuals, I could surmise with high probability regarding those who have traveled widely and those confined to their state or region.

I could not conclude this chapter without discussing personal change. In chapter 12, I commented on organizational change. Change is inevitable in life. The common cliché is the only thing constant is change. Change applies to individuals as it does to organizations. One could argue that an organization's change represents the sum of individuals' change within an organization. Change is personal and organic. Some people glide effortlessly through whatever hand they are dealt in life, while others struggle with the minutest change. It all depends on how one views and handles change and the degree of uncertainty acceptance. Management scholars postulate that uncertainty remains the most fearful element, not change itself, in a change process.

For the newcomer or the native-born, adopting a new mindset is threatening and challenging yet necessary to succeed and thrive. It involves the relinquishing of some beliefs, practices, and way of life and embracing new ones. In managing personal change, nothing illustrates this dynamic and struggle better than the saying by Reinhold Niebuhr, and adopted by Alcoholics Anonymous, often called the serenity prayer, which says: "God, grant me the serenity to accept the things I cannot change, the courage to change the things I can, and the wisdom to know the difference."

Burrowing into this seemingly reasonable approach to managing and navigating personal change and relating it to the discourse in this book; Accepting the things that you cannot change, as a newcomer, you cannot change the culture, norms, and practices of your new land. As a native-born, you cannot compel or force newcomers to return to their native lands. They are here now. For the things one can change, namely: your behavior, attitudes, and understanding of your circumstance, one must be intentional and purposeful to effect the change desired. The wisdom of knowing the difference between what to accept and what to change lies in what is within one's circle of influence and control. It becomes fruitless to fret about what you cannot control. Stephen Covey effectively utilized his circle of concern and influence concept to drive this point home. He advised by focusing energy on those things that you can influence; you can make effective changes. He termed those who focus on what they can influence; proactive people and those who focus on things beyond their control; reactive people.

The learning here is that the level and speed of the newcomer's acculturation and assimilation depends on their propensity for personal change. Likewise, the native-born must possess or acquire the capacity to understand and appreciate the newcomer. Some readers may perceive the suggested attitudinal shifts in assessing and evaluating newcomers as unneeded burden. While this stance might be partially correct, consider the alternative of losing out on the talent battle among organizations. Besides, the mix of the workforce pool is changing. Knowledge and understanding of the population dynamics, and cultural components provide an edge to forward-looking organizations.

"We all smile in the same language."
-George Carlin

CHAPTER 18

CONCLUSION

As you have read, this book contains experiences and stories to illustrate our society's stark realities. Because it is written for self-help, it dwelt more heavily on negative behaviors. This skewing is intended to demonstrate with examples the guts of the newcomers' encounters and occurrences as they grapple with acculturation in their new land. It exposes the native-born to the stresses and tensions the newcomer undergoes. Also, it teases some behaviors of the native-born that are hurtful to the new entrants. These antics may be innocently perpetuated or deliberately unleashed. Nonetheless, once the genie is out of the bottle, you cannot put it back. Remember what the bible says about how our utterances and acts can hurt or nourish (Proverbs 11:17; Proverbs 15:4, and Ephesians 4:29). The harm dished by these acts, comments, and activities persist and sometimes breeds negative counter behaviors.

The newcomer is not absorbed in the responsibility of assimilating into their new land. An equal onus lies on the new entrant to speedily understand and appreciate their new land's culture and norms. No one can perform this inevitable acculturation process for the newcomer. The earlier one engages with their new country; the better one ascends to a higher integration level. There is generally a correlation between the degree of immersion and attained assimilation level. Any tepidness and delay in embracing your new land results in a prolonged and perhaps a more painful acculturation.

In writing this book, the author pondered heavily about not sounding negative and guilt-evoking. In the end, the author considered it wise to share the stories and occurrences. Ignoring them would be helping to maintain behaviors many deem inappropriate, demeaning, and plain ugly. Should you think for a moment, the chances are that you have observed some of these behaviors meted to people around you. You may even have knowingly or unknowingly engaged in some of these acts. The goal here is not to whip one into a remorse frenzy. Instead, to perhaps trigger some introspection, with a view of bettering oneself.

Allow me to reiterate what I have layered throughout this book. When I refer to newcomers, they have legally immigrated to our country with proper governmental approvals and blessings. History has proven that the acceptance of newcomers has been good for America. Imagine the following that I mentioned earlier in this book. In the past two decades, about 25 percent of all venture capital startups were launched by immigrants, including eBay, Google, and Yahoo. Forty percent of recent Silicon Valley innovations originated from immigrants. According to a 2011 report by Forbes, 40 percent of all fortune 500 companies were started by immigrants or their children. These are data that cannot be ignored or discarded. These facts, among others, make America exceptional and the envy of the world.

Some may argue, so what? Should we be jumping for joy then? Not so quick to dismiss, whatsoever. We cannot conveniently cherry pick our likes and dislikes about our great nation. What makes America unique lies in her "melting-pot" persona and ethos. We all coexist to lift a collective lot of our country. Some effortlessly set their eyes on the national prize and work towards its achievement, while others begrudgingly, kicking and screaming pursue the same mission of nation upliftment. Such is the beauty of our democratic country.

The newcomers cannot be passive and pampered into their new society. Far from it. They must work hard to learn and assimilate into their new land. They must be diligent in accessing all available resources and opportunities in their integration journey. A few bad knocks cannot paralyze them along the way. They must be resilient and persevere against all odds. These are the only successful pathways to achieving their American dream.

Numerous studies have validated that diversity and inclusion are advantageous to organizations. According to a McKinsey corporation report, companies with significantly more racial and ethnic diversity are 35 percent more likely to outperform competitors. The same report stated that organizations with racial, ethnic, and gender diversity have a 25 percent greater likelihood of being more profitable. If diversity and inclusion are good for organizations, they should be good for our country as well. Diversity and inclusion include diverse ideas, thoughts, and perspectives in viewing items and subjects. The environment rich in varied viewpoints produces superior solutions and remedies.

Whereas diversity and inclusion are embraced by many, their implementation poses challenges. The oft-reluctance and resistance in deployment stem from the culture that opposes the special treatment of underrepresented groups. This school of thought advances that equality is best attained by eliminating differences in treatment. Diversity and inclusion emphasis elicits defensiveness and obstructionism by other groups. Opponents submit that diversity and inclusion are discriminatory, and therefore, counter-productive. For example, in college admissions, such filtering based on race and ethnicity is harmful, they claim.

Furthermore, they protest that it violates the American creed …… "without regard to race, creed, or color." Those that counter this perspective advance that all groups did not start the proverbial race from the same point. Centuries of disenfranchisement, denials, and suppression leave certain groups behind, and therefore, must be brought to par with the rest of society. Nonetheless, numerous studies have validated that diversity and inclusion are needed elixir in society.

Cultural awareness and sensitivity are not only vital but imperative in society. According to the dictionary definition, cultural awareness refers to the knowledge, awareness, and acceptance of other cultures and others' cultural identities. It reflects the understanding of the differences between themselves and others, especially in attitudes, values, and norms. A culturally aware individual recognizes the nuances and sensibilities of one's and other cultures. Being culturally intelligent does not rob us of our cultural grounding. Instead, it provides the capacity to understand and appreciate individuals with a different worldview fashioned by their background. You can be who you are while recognizing and respecting other people's values. As a business acquaintance once put it, "It does not steal from you but adds much to you."

Behaviorists have advanced the various stages of assimilation in one's new land. They have submitted that the integration process involves both emotional and psychological components. I will attempt to marry their tenets with the various chapters and characters in this book. The first stage is what I refer to as the "awe-struck" phase. This stage is when the newcomer is wowed and excited at their new observations and exposures. Even the simplest things are new and interesting. Remember Cecelia's marvel at the motorists' complete adherence to the traffic signals and lights. Or, Tabita's wonderment at the plentifulness and orderliness of items at a Walmart Super Store. How about Bolade, who could not believe his eyes and ears when he returned a pair of shoes and obtained a full refund without interrogation by the store cashier? Or, the excited Greek family, who could supersize their drinks and obtain refills free. At this "awe-struck" phase, you are consumed by, inundated with, and mesmerized by new awakenings and discoveries. The sheer volume of new things interferes with proper filtration. They come at you, and you are haplessly taking them in.

The newcomer then slips into the "reality" stage, when these initially wonderful features and discoveries dim in astonishment. You are now immersing in your new land and encountering the harsh realities of your new society — the paradox of America. This phase is usually marked by frustration, feelings of discontent, and inadequacy. Remember the many "Houston, we have a problem" episodes in the various chapters. Ben and his English Language proficiency minimization to the Arizona schoolmates, who could not secure a hotel room for their lack of a credit card. How about soccer mom Anna, who received a dose of reality from her fellow soccer moms? Do not forget Jesse and his fiancée on their new home and neighborhood experience.

At this juncture, the newcomer wonders if they can successfully navigate the curves inherent in the new land's culture and norms. It is here that some mentally determine to engage, while others are ambivalent or decide to delay the embrace of their new land's culture. They often pay a heavy price because they become laggards in acculturation and miss out on things. For example, remember Andrew, who refused to indulge in the American credit system and its creditworthiness establishment for many years. He confessed that his delay in building this essential financial viability cost him dearly in later years when needed.

The newcomer glides to the learning, rationalization, and adjusting phase. They start to self-educate on cultural relativism. Their earlier feelings of incompetence in handling the myriad challenges confronting them turn to acceptable balance to the hand dealt them. This comfort stems from better familiarity and understanding of the new land's culture and practices. They have now determined to engage in their new land, hence my term of the "Plunging" phase.

With the resolve to commit, they see and feel the more positive aspects of their new land. They become more tolerant of some unavoidable knocks along the way. They generally steel their backbones and wholeheartedly go for it. They start to morph into what I discussed in chapter 16 as "hybrid self." The tensions between your old and new lands' cultures magnify and multiply. The cultural conflicts intensify as they shed most of the old country's ways and embrace the new country's culture. Observe that I do not imply a total abandonment of their old country's culture or an absolute embrace of their new land's norms and practices. Such would represent a total surrender and, therefore, impossible to attain. You could not accomplish this even if you tried. The futility of such an attempt is because there remain certain aspects of who we are that have already been chiseled in us. We can never deconstruct them.

While transitioning into their "hybrid Selves," they undertake trips to their old country. These visits reinforce their decision to emigrate, and ironically, compels them to appreciate their new country more. I call this the "Affirmation" stage. They are availed of a comparative society, America, that amplifies the brokenness of their old country. These drawbacks were not readily seen or felt while living in the old country. They were simply a part of the norm in those societies. The newcomer hurriedly returns and positively fine tunes their assimilation journey.

Now that you are almost done reading this book, I venture to provoke some thoughts. I am hereby asking for a waiver. Here we go .....

Should you find yourself easily dismissive of many or all of the tenets of this book, I urge you to reconsider. Should you even find yourself angry, not at the perpetrators of some of these intolerant behaviors, but on the sheer fact that they are being called out, I would plead with you to refocus the mirror to the proper angle. Should you be quick first to point accusing fingers at someone else for practicing these behaviors without searching your soul first, may you please pray about it. Should you be mired in the mindset of "they are here if they don't like it, get a one-way ticket home," may you find it in your heart to understand that the newcomer is here now and possibly entrenched. Of course, legally. Should the urge of defensiveness, the twin sister of intolerance rooted in race and ethnicity, creep upon you, please work to resist it. It is usually easier to rationalize our actions and behaviors than confronting them head-on. Recognizing our challenges is the first critical step towards their resolutions.

In the good book, we are told to "…. love thy neighbor as thyself." (Leviticus 19:18). The Apostle Luke wrote how a law expert attempted to test Jesus by asking who our neighbor is (Luke 10:29). Jesus flipped it on him by telling the story of the Good Samaritan. Remember, the Samaritans were not on equal footing with the rest of the populace. They saw them as lesser. Jesus helped the law expert to understand who our neighbors truly are. It could be the Samaritan whom the populace had less regard for. I am not equating newcomers to the Samaritans by any means. The purpose of this reference is to illustrate that our "neighbors" are around us. Jesus, through this parable, taught loving your neighbor goes beyond only loving those in our neighborhoods. It is larger than that. Neighbors could be our coworkers, next-door neighbors, school and college mates, fellow soccer moms, fellow NASCAR racing fans, newcomers to our land, etc. We live in the same community as our neighbors. We are all inextricably intertwined, whether we want or not.

As humans, we are imperfect and possess flaws and biases. Some of these biases are unconscious and, therefore, fly under our conscious radar. Unfortunately, some of our actions, including inappropriate and hurtful ones, are driven by these tendencies. They exist. Therefore, being aware of them alerts us to be vigilant to avoid their subjugation. In some instances, many have recalibrated their worldview based on more awareness of their emotions, proclivities, and biases.

May you continue to make our community a better place.

GLOSSARY

African proverb: African sayings held as wisdom from ancestors

Amygdala hijack: Immediate emotional response triggered by the limbic system

Arranged marriage: Marital union with couple primarily selected by individuals other than the couple themselves

Ashanti: A tribe in Ghana, West Africa

Atta boy: Encouragement and positive affirmation

AWOL: Absent without leave, notice, or approval

Banquet of cultures: Description of America

Biculturalism: Co-existence of two distinct cultures

Black night club: Entertainment venue patronized primarily by African American goers

Bolade: African name

Bona fide: Genuine and real

Bosnia and Herzegovina: A country in the Balkans (Southeast Europe)

Clunker: An old, run-down vehicle or machine

COVID-19 Pandemic: Virus that infected the world in 2020/2021

Cultural awareness and sensitivity: Understanding of the differences between oneself and people from other countries and backgrounds

Cultural intelligence: Capacity to relate and work effectively across cultures

Cultural relativism: Relating culture to its environment and standards

Drive-in movie theater: Outdoor cinema with large screen, projection booth, and parking area for automobiles

Dude: Another name for man

Elbow room: Adequate space for a person to operate in

Eskimo: Inhabitants of the Arctic region consisting of the Inuits, Yupiks, and Aleuts tribes

Ethnic food: Food originating from an ethnic group utilizing knowledge of local plants and animals

Ethnic store: Store that sells food items that are distinctly cultural

Ethnocentrism: Regarding one's culture as the correct way of living

Ethos: Character

Faux pas: An embarrassing act or remark in a social situation

Foie gras: Specialty food made of duck or goose liver

Free ride: A benefit obtained without the usual cost or effort

Giving the finger: Obscene hand gesture communicating strong contempt

Hiking: Walking in nature as recreation

Hippie: A member of the counterculture of the 1960s

HIV: Human immunodeficiency virus that attacks the body

Jaywalking: Cross or walk in the street unlawfully or without regard for approaching traffic

Jekwu: African name

Kaleidoscope of cultures: Description of America

Magnificent mile in Chicago: Vibrant district with upscale shops in Chicago

Mardi gras: Annual Fat Tuesday carnival celebration (largest crowd in the U.S. is in New Orleans

Melting pot: Description of America

Milking: Taking undue advantage
Moonlight: Second job in addition to one's regular employment
Mozambique: A southern African country
Native tongue: Language exposed to from birth or ethnic language
New arrival: New immigrant
Newcomer: New immigrant
New entrant: New immigrant
No-no: Not possible or acceptable
Obesity: Abnormal and excessive fat accumulation. Body Mass Index (BMI) over 30
Obsequies: Funeral or burial rites
Oops: Error sound
Ouch: Distressing sound
Oxford English: The British King's English, usually regarded highly
Remote working: Working style that allows work outside of a traditional office environment, such as homes, hotels, coffee shops
Restroom: Another name for toilet and lavatory
Right to disconnect law: French law that employers shall not encroach on employees' personal lives with calls and emails
Rodeo drive in Los Angeles: Vibrant district with upscale shops in Los Angeles
Rwanda: A country in East Africa
Shacking up: Cohabitating as unmarried partners
Shark: Predator or attacker, usually in business or workplace
Séance: A spiritual meeting to receive spirit communications
Shaman: Religious or mystical expert who functions as a healer, prophet, and custodian of culture, and communicates with the spirits

Soccer mom: Middle-class suburban mother who actively engages and shuttles her children to play soccer and other activities

Sugar daddy: A rich older man who lavishes gifts on a young woman in return for her company or sexual favors

Summa Cum Laude: Graduating with the highest distinction

Supersizing: Greatly increasing size of an item, usually food or drinks

Sushi: Sour-tasting combination of rice and either raw or cooked fish

State Security Service: Government agency tasked with nation's security in some countries

Tattoo: A mark, figure, design or word placed on the skin

Tipping: Gratuity given for service provided in addition to the basic price of the service

Traditional: Customary, long-established, and past practice

Trouser: Another name for a pair of pants

Unconscious bias: Social stereotypes formed outside one's conscious radar

Weed: Another name for marijuana

Woodstock music festival: Music festival held in 1969 near Woodstock, New York during the "Peace Movement" era

Zantac: Medication to treat heartburn

Zimbabwe: A southern African country

2nd Amendment: Right to keep and bear arms in America

CPSIA information can be obtained
at www.ICGtesting.com
Printed in the USA
BVHW061621250122
627124BV00013B/477